Plasma Technology for Biomedical Applications

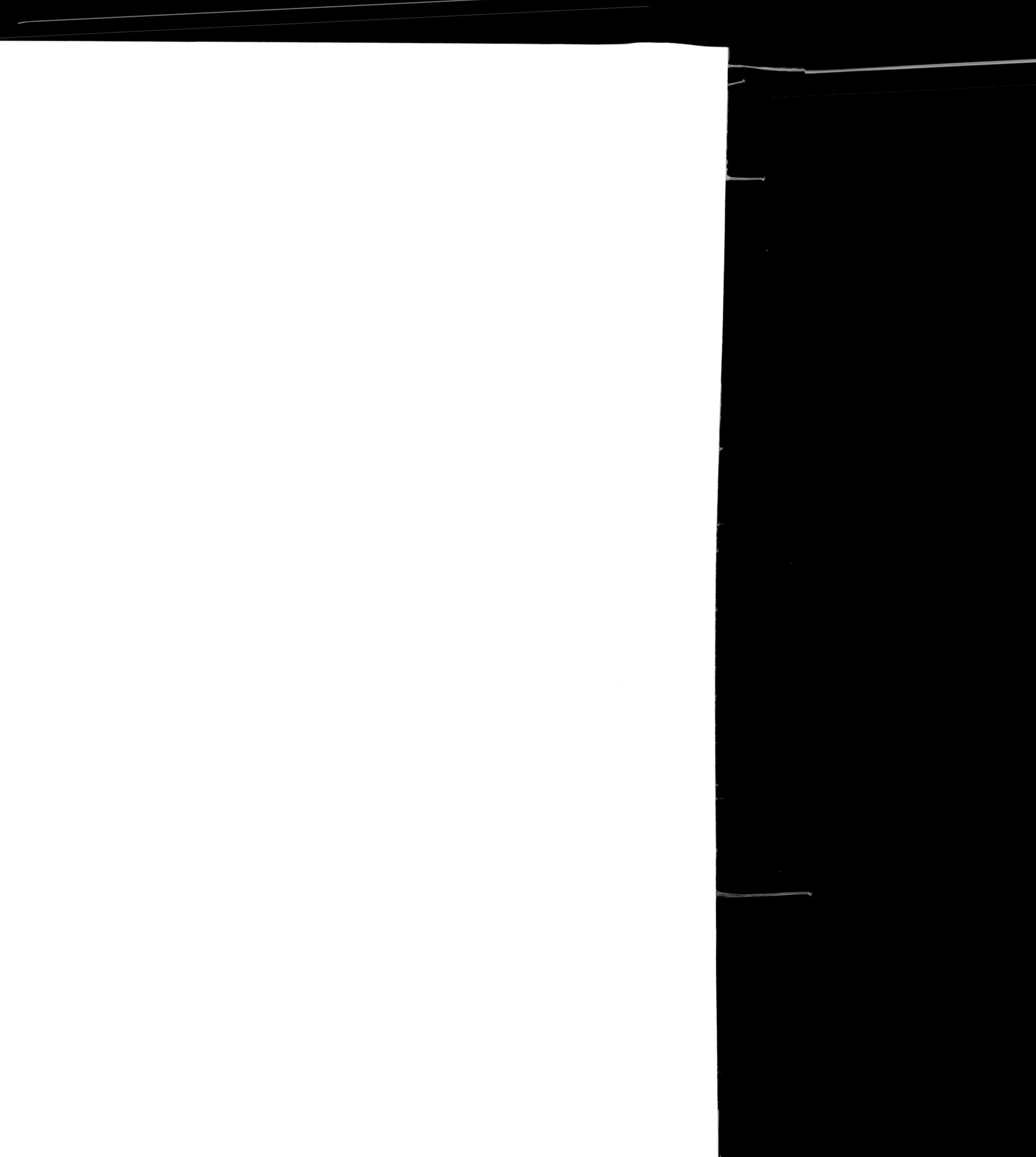

Plasma Technology for Biomedical Applications

Special Issue Editor

Emilio Martines

MDPI • Basel • Beijing • Wuhan • Barcelona • Belgrade

Special Issue Editor
Emilio Martines
Consorzio RFX
Italy

Editorial Office
MDPI
St. Alban-Anlage 66
4052 Basel, Switzerland

This is a reprint of articles from the Special Issue published online in the open access journal *Applied Sciences* (ISSN 2076-3417) from 2019 to 2020 (available at: https://www.mdpi.com/journal/applsci/special_issues/Plasma_Biomedical_Applications).

For citation purposes, cite each article independently as indicated on the article page online and as indicated below:

LastName, A.A.; LastName, B.B.; LastName, C.C. Article Title. *Journal Name* **Year**, *Article Number*, Page Range.

ISBN 978-3-03928-736-9 (Pbk)
ISBN 978-3-03928-737-6 (PDF)

Cover image courtesy of Gianluca De Masi.

Contents

About the Special Issue Editor

Emilio Martines is a plasma scientist. After earning a degree in Physics at the University of Pisa (Italy) in 1991 he moved to Padova (Italy), to join Consorzio RFX, a research institution involved in the study of magnetized plasmas for controlled thermonuclear fusion research. There he earned a Ph.D. in Energetics and then was hired by the National Research Council (CNR), to work within Consorzio RFX. In 2007, he was appointed as the leader of one of the research groups within the institution. During his scientific career, he has studied many different issues, including turbulence at the edge of fusion devices, helical equilibria in reversed field pinch plasmas, and different kinds of industrial plasma applications. For ten years, he has been active in the field of plasma medicine, leading and developing this research line within Consorzio RFX. He is also a professor of Nuclear Fusion and Plasma Applications at the University of Padova. He is the co-author of more than 160 papers published in international journals, with an H-index of 33, and several patents. He is a member of the International Society of Plasma Medicine.

Editorial

Special Issue "Plasma Technology for Biomedical Applications"

Emilio Martines [1,2]

[1] Consorzio RFX, 35127 Padova, Italy; emilio.martines@igi.cnr.it
[2] Istituto per la Scienza e Tecnologia dei Plasmi del CNR, 35127 Padova, Italy

Received: 18 February 2020; Accepted: 19 February 2020; Published: 24 February 2020

1. Introduction

The use of plasmas for biomedical applications in encountering a growing interest, especially in the framework of so-called "plasma medicine", which aims at exploiting the action of low-power, atmospheric pressure plasmas for therapeutic purposes [1–4]. Several applications have already reached the stage of clinical trials, while others are on their way, a large set of different plasma sources able to work at atmospheric pressure with low dissipated power have been created, and some of them are already certified as medical devices. From the scientific viewpoint, action mechanisms for the interaction of plasmas with cells, tissues and pathogens are being elucidated, although this is a slower process which still requires great efforts. Furthermore, the indirect action through the use of plasma-treated liquids is also being explored, presenting promising possibilities. Finally, one should mention the possibility of plasma-cell interactions not directly related to a therapeutic action of the plasma, but of great importance for facilitating other therapeutic approaches, such as plasma-mediated gene transfection and drug penetration.

The plasmas used in this kind of applications have two main requirements: To be produced at atmospheric pressure, and to keep the treated substrate at temperatures below 37 °C. These two requirements imply that we are dealing with cold atmospheric plasmas (CAP), where only the electrons have a high temperature (of the order of 1 eV, that is 11,600 K), while ions and the neutral molecules are at or near room temperature. These are weakly ionized plasmas, where most of the gas molecules are neutral, and electron-neutral collisions are the main drive of transport processes. In order to keep the power deposition, and thus the gas heating, to low values, a method for limiting the current needs to be employed in the plasma generation, so as to avoid transition to an electric arc: The two most widespread approaches are the dielectric barrier discharge (DBD), where a dielectric layer separating the electrodes rapidly extinguishes the current when enough charge deposits on it, and the radiofrequency (RF) discharge, where the voltage is reversed very fast, at frequencies above 1 MHz [5].

This special issue was launched to collect the latest advancements in this exciting and interdisciplinary field of research. There were 13 papers submitted, of which 11 papers were accepted. When looking back to this special issue, various topics have been addressed: Mechanisms of interaction of plasma with substrate (two papers), technologies for production of plasma-activated water (two papers), and applications to cancer treatment (three papers), disinfection (two papers), regenerative medicine (one paper) and dentistry (one paper).

2. Interaction of Plasma with Substrate

There are two papers in this special issue dealing with the problem of the interaction of the plasma with the substrate, and in particular with substrates composed of living tissues. The first one, by Schweigert and co-workers, deals with a problem which has gained considerable interest in the last few years, that is the effect on the plasma-substrate interaction of the substrate grounding condition [6]. The study was performed both experimentally, using DBD sources operating in helium and argon with cylindrical and planar geometries to treat cancer cells in vitro, and through 2D simulations. It was shown that a metal grounded target positioned below the plate with cells and medium led to an increase of the electric field over the plasma-medium interface, resulting in higher electron energy and density and OH-radical production rate. As a consequence, the ability of killing cancer cells was enhanced, pointing to the importance of grounding to achieve relevant biological effects.

In the second paper, Cordaro and co-workers analyzed the performance of a plasma source, based on the helium DBD jet concept, designed for non-thermal blood coagulation [7]. They demonstrated that the plasma action indeed accelerates platelet aggregation and fibrin formation, thus inducing coagulation, while at higher powers and for longer treatment times also harmful effects appear, such as red cells lysis, with destruction of collagen fibers and dehydration of muscle fibers. In carrying out this task, it was ascertained that the main sample heating mechanism was due to the electric current flowing to the sample. This led to the conclusion that a power deposition evaluation performed on sample targets (as could be prescribed, for example, in a technical norm) could not be representative of what happens when the plasma is applied to actual living tissues. In agreement with the previous paper, this also implies that the grounding condition of the substrate is an important issue.

3. Production of Plasma-Activated Water

The indirect treatment realized using water or other liquids (most often, cell culture medium) previously treated with the plasma is an important part of plasma medicine studies. A wealth of devices have been proposed to perform this treatment in an efficient way [8].

In the first paper of this section, a new device based on a low-current arc formed in ambient air is proposed [9]. The idea (already exploited also by other authors) is that, being the treatment indirect, the requirement of low thermal load holding for direct plasma treatments can be relaxed, so that the plasma used to treat liquids does not need to strictly be a CAP. The authors demonstrated the possibility of generating a stable discharge, even when using liquids that have low electrical conductivity, and confirmed the possibility of treating a continuously flowing liquid. The concentration of reactive oxygen and nitrogen species in water after treatment using the low-current arc object of this study was two orders of magnitude higher than that of water treated using conventional CAP under similar conditions. Strong bactericidal effects of the treated water were demonstrated on *Escherichia coli* cultures.

The second paper, by Schmidt and co-workers, uses a different approach, that is an inductively limited discharge, to treat large volumes of water, overcoming the limitation of few millilitres associated to the most usual approaches [10]. The proposed technique uses high voltage leakage transformers for discharge current limitation to avoid arcing. The authors treated tap water, saline solution and distilled water, and observed that, except for tap water, the treated liquids became acidic and the conductivity increased. In all liquids, distinct nitrification was observed. The microbiological studies showed that physiological saline solution and tap water became antimicrobial. The authors concluded that with the proposed, portable setup, significant volumes of plasma-treated water can be easily produced in less than 30 min.

4. Cancer Treatment

There are several intriguing evidences that the reactive oxygen species (ROS) originated in the plasma can kill cancer cells selectively, preserving healthy ones. This has been attributed, in analogy to other redox therapies, to the faster formation and loss level of these species in cancer cells as compared to healthy ones, leading to higher baseline concentrations. The plasma action will raise ROS concentration in both cell types, but only in cancer cells this will go beyond the threshold leading to apoptosis [11]. Reactive nitrogen species are also considered to be important in determining the plasma action, although their role is less understood [12].

In the paper of Hasse and co-workers, a pre-clinical study on the use of argon plasma as adjuvant therapy on progressive head and neck cancer is described [13]. The plasma was produced by the certified medical device kINPen MED, operating at radiofrequency with an argon flow. The response of healthy and tumour cells of head and neck cancer to CAP exposure was addressed. In addition, tissue samples from 10 patients with histologically proven cancers of the maxillofacial region were treated and then investigated for induction of apoptosis and secreted proteins with antitumour activity. The viability of cancer cells was found to be strongly reduced by the plasma action, although no clear selectivity of cancer cells could be observed. However, induction of apoptosis was superior in tumour tissue than in healthy mucosal tissue. Furthermore, CAP treatment significantly decreased cell motility in squamous cell carcinoma cells only but not in non-malignant keratinocytes. Overall, these results point to CAP treatment as a promising adjuvant treatment option to eliminate minimal residual cancer cells after radical surgery of carcinoma.

The antitumour effect of the reactive species generated by the plasma can be displayed through direct plasma application, as in the previous study, or through the application of liquids previously treated with the plasma. This is the theme of the second paper of this section, by Nguyen and co-workers, which shows that plasma-activated medium can be stored for up to six months in a freezer and then exhibit cytocidal effects on human cervical cancer HeLa cells, similar to those of a direct plasma treatment [14]. The treated medium was Dulbecco's Modified Eagle Medium supplemented with 10% fetal bovine serum and antibiotics, and the treatment was performed with a micro plasma-jet nozzle operating in air flow. The cytocidal effect was attributed to H_2O_2 and nitrite/nitrate formed in the liquid by the plasma action.

Apart from the previoulsy mentioned mechanism, a new paradigm is also taking momentum as a possible way of inducing cancer cell death through the plasma action. This is the immunogenic cancer cell death, which proposes stimulation of an immune response against apoptotic tumor cells, and is the theme of the third paper, by Rodder and co-workers [15]. In this work, the role of plasma-treated murine melanoma cells in modulating murine immune cells' activation and marker profile was investigated. The results indicate a tumor-static action in terms of metabolic activity and cell motility and a negligible protective effect of protein present during the treatment. A role of plasma-mediated activation of splenic immune cells and a modulation of inflammatory parameters, in agreement with a pro-immunogenic role of plasma treatment, were also observed.

5. Disinfection

The ability of atmospheric pressure plasma to kill bacteria, either by disruption of the cell envelope or through more subtle effects [16], is well known, and indeed is the first effect which has been invoked for utilization in medical practice [17].

The first paper of the special issue related to CAP disinfection properties is a review, by Gupta and co-workers, concerning the effectiveness of this technique on biofilms [18]. While most studies in this field test the plasma action against bacteria cultures in planktonic form, in real life applications bacteria are often found in the form of biofilms, that is groups of microorganisms adhered to a substrate within a self-produced matrix of extracellular polymeric substance, mostly composed of water, polysaccharides, proteins, and extracellular DNA. Biofilms are much more resilient to antibiotics and antiseptics, thanks to the fact that the extracellular polymeric substance forms a physical barrier, responsible for limiting the transport of chemicals into and out of the biofilm, and are usually challenging to eradicate. The ability to significantly affect biofilms is thus a crucial property to be demonstrated if plasma-based disinfection is to be brought to the market. The paper of Gupta et al. reviews the existing literature on biofilm eradication through CAP, concluding that the technology appears to be promising, but further effort is required in developing (and certifying, I would add) plasma sources adequate for use in real world environments.

The second paper, by Liu and co-workers, is aimed at fully understanding the bio-decontamination process in a reduced-pressure oxygen plasma, using *Escherichia coli* as the target microorganism [19]. This study does not completely fall into the plasma medicine realm, as defined in the introduction, in the sense that it makes use of low-pressure conditions. It is however very interesting, as it makes a thorough comparison of the role of different agents, that is UV radiation, charged species and free radicals, in the decontamination process, and thus offers valuable information also for processes operating at atmospheric pressure. In particular, the authors found that the essential effectiveness on *E. coli* of the oxygen plasma can to be attributed to the intense etching action of charged species, that is electrons and ions, on the bacilli materials. Lipid peroxidation in the cell membrane by oxygen radicals plays a major role only during the initial phase (< 40 s), and is then restrained by the effect of charged particles. The function of UV radiation is to assist in the whole process, resulting in slight damage and rupture of DNA. This study, while confirming the marginal role of UV radiation, adds another bit to the ongoing debate about the importance of charged species in the deactivation process.

6. Regenerative Medicine

The beneficial effects of plasma exposure in regenerative processes, such as wound healing, is one of the most advanced applications of plasma medicine. Holganza and co-workers have studied the effects of the exposure to a helium plasma produced in a DBD plasma jet on tadpoles of *Xenopus laevis*, in relation to developmental effects such as tail regeneration and metamorphosis [20]. The effect of plasma treatment following tail amputation was investigated. The experiment confirmed previous observations about the fact that the plasma treatment accelerates tail regeneration while slowing down the metamorphic progress, the latter possibly indicating the physiological cost of enhanced regeneration involving metabolic machinery at the cellular and organelle level. These effects, associated to higher oxidative stress, were linked to increase in Ca^{2+} content during wound healing, possibly derived from extracellular stores such as the endoplasmatic reticulum. Additionally, adherens junctions between epidermal cells of the tail and reduction of intercellular spaces following plasma exposure were observed, indicating adaptive changes in order to maintain skin integrity.

7. Dentistry

The paper by Shahmohammadi Beni and co-workers [21] adds to the application for which the first plasma medicine tool was originally designed [17], that is the use of a CAP to perform dental treatments in the oral cavity [22,23]. The authors have investigated numerically the transport by convection and diffusion of OH radicals and of hydrogen peroxide (H_2O_2) generated by CAP over treated teeth. This is important, as OH radical and hydrogen peroxide are two of the most important reactive species generated by the plasma, in terms of biological effects. The model used by the authors consists of an equation for the carrier gas motion, based on the level set method, that is a conceptual framework allowing to perform numerical computations involving curves and surfaces on a fixed Cartesian grid without having to parametrize these objects, and transport equations giving the evolution of the OH radical concentration and of the H_2O_2 concentration. The equations were solved in a realistic geometry model of the mandibular jaw and of the space between it and the plasma source. The simulation results allowed a realistic evaluation of the deposition of the two active species on the different teeth of the simulated jaw. Overall, apart from the scientific merit of the specific results, this code appears to be a valuable tool for carefully assessing the actual deposition of active chemical species (possibly including also other species) in a complex geometry, thus allowing to optimize the design of the plasma source, and could be possibly extended also to different environments and plasma treatments.

8. Conclusions

As a matter of fact, it is clear that the strongly interdisciplinary plasma medicine community is evolving towards a higher level of scientific depth and analysis detail, while at the same time progressing towards bringing applications from the laboratory to the patient. This is crucial to fulfill the expectations created by this new discipline, which is foreseen to soon become, at least in some contexts, part of the tools routinely available to the practitioner.

Funding: This editorial received no external funding.

Acknowledgments: Thanks are due to all the authors and peer reviewers for their valuable contributions to this Special Issue. The MDPI management and staff are also to be congratulated for their untiring editorial support for the success of this project.

Conflicts of Interest: The author declares no conflict of interest.

References

1. Fridman, G.; Firedman, G.; Gutsol, A.; Shekhter, A.B.; Vasilets, V.N.; Fridman, A. Applied plasma medicine. *Plasma Process. Polym.* **2008**, *5*, 503. [CrossRef]
2. Kong; Kroesen, M.G.; Morfill, G.; Nosenko, G.T.; Shimizu, T.; van Dijk, J.; Zimmermann, J.L. PLasma medicine: An introductory review. *New J. Phys.* **2009**, *11*, 115012. [CrossRef]
3. von Woedtke, T.; Reuter, S.; Masur, K.; Weltmann, K.D. Plasmas for medicine *Phys. Rep.* **2013**, *530*, 291. [CrossRef]
4. Graves, D.B. Low temperature plasma biomedicine: A tutorial review. *Phys. Plasmas* **2014**, *21*, 080901. [CrossRef]
5. Weltmann, K.-D.; Kindel, E.; von Woedtke, T.; Hähnel, M.; Stieber, M.; Brandenburg, R. Atmospheric-pressure plasma sources: Prospective tools for plasma medicine. *Pure Appl. Chem.* **2010**, *82*, 1223. [CrossRef]
6. Schweigert, I.; Zakrevsky, D.; Gugin, P.; Yelak, E.; Golubitskaya, E.; Troitskaya, O.; Koval, O. Interaction of Cold Atmospheric Argon and Helium Plasma Jets with Bio-Target with Grounded Substrate Beneath. *Appl. Sci.* **2019**, *9*, 4528. [CrossRef]
7. Cordaro, L.; De Masi, G.; Fassina, A.; Gareri, C.; Pimazzoni, A.; Desideri, D.; Indolfi, C.; Martines, E. The Role of Thermal Effects in Plasma Medical Applications: Biological and Calorimetric Analysis. *Appl. Sci.* **2019**, *9*, 5560. [CrossRef]
8. Bruggeman, P.J.; Kushner, M.J.; Locke, B.R.; Gardeniers, J.G.E.; Graham, W.G.; Graves, D.B.; Hofman-Caris, R.C.H.M.; Maric, D.; Reid, J.P.; Ceriani, E.; et al. Plasma-liquid interactions: A review and roadmap. *Plasma Sources Sci. Technol.* **2016**, *25*, 053002. [CrossRef]
9. Gamaleev, V.; Iwata, N.; Hori, M.; Hiramatsu, M.; Ito, M. Direct Treatment of Liquids Using Low-Current Arc in Ambient Air for Biomedical Applications. *Appl. Sci.* **2019**, *9*, 3505. [CrossRef]
10. Schmidt, M.; Hahn, V.; Altrock, B.; Gerling, T.; Gerber, I.C.; Weltmann, K.-D.; von Woedtke, T. Plasma-Activation of Larger Liquid Volumes by an Inductively-Limited Discharge for Antimicrobial Purposes. *Appl. Sci.* **2019**, *9*, 2150. [CrossRef]
11. Graves, D.B. The emerging role of reactive oxygen andnitrogen species in redox biology andsome implications for plasma applicationsto medicine and biology. *J. Phys. D Appl. Phys.* **2012**, *45*, 263001. [CrossRef]
12. Semmler, M.L.; Bekeschus, S.; Schafer, M.; Bernhardt, T.; Fischer, T.; Witzke, K.; Seebauer, C.; Rebl, H.; Grambow, E.; Vollmar, B.; et al. Molecular mechanisms of the efficacy of cold atmospheric pressure plasma (CAP) in cancer treatment. *Cancers* **2020**, *12*, 269. [CrossRef]
13. Hasse, S.; Seebauer, C.; Wende, K.; Schmidt, A.; Metelmann, H.-R.; von Woedtke, T.; Bekeschus, S. Cold Argon Plasma as Adjuvant Tumour Therapy on Progressive Head and Neck Cancer: A Preclinical Study. *Appl. Sci.* **2019**, *9*, 2061. [CrossRef]
14. Nguyen, N.H.; Park, H.J.; Hwang, S.Y.; Lee, J.-S.; Yang, S.S. Anticancer Efficacy of Long-Term Stored Plasma-Activated Medium. *Appl. Sci.* **2019**, *9*, 801. [CrossRef]
15. Rödder, K.; Moritz, J.; Miller, V.; Weltmann, K.-D.; Metelmann, H.-R.; Gandhirajan, R.; Bekeschus, S. Activation of Murine Immune Cells upon Co-culture with Plasma-treated B16F10 Melanoma Cells. *Appl. Sci.* **2019**, *9*, 660. [CrossRef]
16. Martines, E. Interaction of cold atmospheric plasmas with cell membranes in plasma medicine studies. *Jpn. J. Appl. Phys.* **2020**, *59*, SA0803. [CrossRef]
17. Stoffels, E.; Flikweert, A.J.; Stoffels, W.W.; Kroesen, G.M. Plasma needle: A non-destructive atmospheric plasma source for fine surface treatment of (bio) materials. *Plasma Sources Sci. Technol.* **2002**, *11*, 383. [CrossRef]
18. Gupta, T.T.; Ayan, H. Application of Non-Thermal Plasma on Biofilm: A Review. *Appl. Sci.* **2019**, *9*, 3548. [CrossRef]

19. Liu, H.; Feng, X.; Ma, X.; Xie, J.; He, C. Dry Bio-Decontamination Process in Reduced-Pressure O2 Plasma. *Appl. Sci.* **2019**, *9*, 1933. [CrossRef]
20. Holganza, M.V.; Rivie, A.; Martus, K.; Menon, J. Modulation of Metamorphic and Regenerative Events by Cold Atmospheric Pressure Plasma Exposure in Tadpoles, Xenopus laevis. *Appl. Sci.* **2019**, *9*, 2860. [CrossRef]
21. Shahmohammadi Beni, M.; Han, W.; Yu, K.N. Dispersion of OH Radicals in Applications Related to Fear-Free Dentistry Using Cold Plasma. *Appl. Sci.* **2019**, *9*, 2119. [CrossRef]
22. Arora, V.; Nikhil, V.; Suri, N.K.; Arora, P. Cold atmospheric plasma (CAP) in dentistry. *Dentistry* **2014**, *4*, 1. [CrossRef]
23. Gherardi, M.; Tonini, R.; Colombo, V. Plasma in dentistry: brief history and current status. *Trends Biotechnol.* **2017**, *36*, 583. [CrossRef]

Article

Interaction of Cold Atmospheric Argon and Helium Plasma Jets with Bio-Target with Grounded Substrate Beneath

Irina Schweigert [1,2,*,†], **Dmitry Zakrevsky** [3,4,†], **Pavel Gugin** [3], **Elena Yelak** [4], **Ekaterina Golubitskaya** [5,6,†], **Olga Troitskaya** [5,†] and **Olga Koval** [5,6,†]

[1] Khristianovich Institute of Theoretical and Applied Mechanics, 630090 Novosibirsk, Russia
[2] George Washington University, Washington, DC 20052, USA
[3] A.V. Rzhanov Institute of Semiconductor Physics, 630090 Novosibirsk, Russia; zakrdm@isp.nsc.ru (D.Z.); gugin@isp.nsc.ru (P.G.)
[4] Department of Physical Engineering, Novosibirsk State Technical University, 630090 Novosibirsk, Russia; Lena.yelak@gmail.com
[5] Institute of Chemical Biology and Fundamental Medicine, 630090 Novosibirsk, Russia; katerinagolubitskaya@gmail.com (E.G.); troitskaya_olga@bk.ru (O.T.); o_koval@ngs.ru (O.K.)
[6] Department of Molecular Biology, Novosibirsk State University, Novosibirsk 630090, Russia
* Correspondence: ischweig@yahoo.com or ivschweigert@email.gwu.edu
† These authors contributed equally to this work.

Received: 4 September 2019; Accepted: 21 October 2019; Published: 25 October 2019

Abstract: The cold atmospheric pressure plasma jet interaction with the bio-target is studied in the plasma experiment, 2D fluid model simulations, and with MTT and iCELLigence assays of the viability of cancer cells. It is shown, for the first time, that the use of the grounded substrate under the media with cells considerably amplifies the effect of plasma cancer cell treatment in vitro. Plasma devices with cylindrical and plane geometries generating cold atmospheric plasma jets are developed and tested. The sequence of the streamers which forms the plasma jet is initiated with a voltage of 2.5–6.5 kV applied with the frequency 40 kHz. We suggest using the grounded substrate under the bio-target during the plasma jet treatment of cancer cells. The analysis of the measured plasma spectra and comparison of OH-line intensity for different voltages and gas flow rates allows us to find a range of optimal plasma parameters for the enhanced OH generation. The time-dependent viability is measured for human cell lines, A431 (skin carcinoma), HEK 293 (kidney embryonic cells), and A549 (human lung adenocarcinoma cells) after the plasma jet treatment. The results with cell-based experiments (direct treatment) performed with various plasma jet parameters confirm the maximum efficiency of the treatment with the optimal plasma parameters.

Keywords: cold atmospheric plasma jet; plasma device; bio-target; plasma-surface interaction

---

1. Introduction

Recently, plasma devices generating the streamer type of breakdown in a mixture of noble gases and air have been widely used in medicine (see, for example [1,2]). The most typical plasma devices operate at 10–50 kHz frequency with the voltage of 2.5 kV–20 kV, applied to the electrode embedded inside of the dielectric tube. The streamer appears over the positive cycle of the applied voltage and propagates inside and outside of the dielectric tube [3,4]. Usually a noble gas is pumping through the dielectric tube since the critical voltage of the breakdown in the noble gases is essentially lower that in the atmospheric air. The streamers propagate over a laminar jet of a noble gas and induces multiple chemical reactions in the mixture of nitrogen, oxygen, water vapor and noble gas some distance apart from the dielectric tube inlet [5,6].

The bio-target treated with the plasma jet is exposed to a chemical cocktail of different radicals, ions and the large electric field delivered by the streamer head. The efficiency of treatment depends on numerous parameters, such as the discharge voltage and frequency, plasma device geometry, type of working gas, velocity of flow gas, distance between a discharge tube nozzle and tissue, time of exposing to the plasma etc.

The effect of different types of targets placed in free plasma jets on the plasma characteristics and OH-production was studied previously. In numerical calculations in Ref. [7], it was shown that an increase of permittivity of targets ranged from plastics to metals enhances the speed of ionization wave and the electron density in the plasma column. In Ref. [8], the images of plasma interaction with different targets showed an increase of intensity of the optical emission for the grounded electrode beneath. The electric field profile between the nozzle and target over the He plasma jet generated with 30 kHz discharge with 2 kV voltage amplitude was measured in Ref. [9], using optical emission spectroscopy on a forbidden line of Helium (2^1P4^1F). The electric field E delivered by the streamer to the metal grounded electrode is shown to be higher than E for the glass target case. A considerable rise in OH density values was found in Ref. [10] with the presence of the metal target compared to the free jet results at the high AC voltage amplitude of 10 or 14 kV. This enlarged value of the OH density was provided with the counter-propagating streamer after impinging the target. In Ref. [11], an original method to increase the OH generation was developed. Using multiple ring electrodes, more OH-radicals were generated. Compared to the case with only one ring, the device with 12 ring electrodes can generate 3–5 times more OH. It was shown that the multiple electrodes enhance the plasma and OH-radical production only inside the tube rather than in the plasma plume in the surrounding air. A higher discharge current and active species densities were achieved in Ref. [12] with the installation of an additional floating inner electrode in the plasma device compared to the device only with two outer electrodes. In Ref. [13], an external biased ring electrode installed between the plasma device and target was used to intensify the streamer characteristics near the target surface. It was shown that the surface electric field and ionization rate were much higher on the dielectric surface than on the conductive one, due to an accumulation of the surface charge, especially with the presence of negatively biased external electrode.

In this work, in the experiment and 2D fluid model simulations, we study the influence of the presence of the grounded substrate beneath the plate with the media and cells on plasma characteristics and the efficiency of cancer cell treatment. Based on the theoretical and experimental results the optimal conditions of CAP jet are formulated and implemented in cancer cell-based experiments. Monitoring the viability of cancer human cell lines A549 and A431 after the CAP jet treatments with various plasma conditions confirms an increase of plasma impact for the optimal plasma parameters.

Following a description of the experimental setups in Section 2, the results of measurements of plasma jet characteristics and the intensity of OH-line in spectra are given in Section 3. A brief description of 2D fluid model and a comparison of simulation results of plasma-target interaction with and without the grounded substrate are presented in Section 4. Materials and methods for study of cells response on CAP treatment are provided in Section 5 and an influence of CAP jet treatment on the viability of A549, A431 and HEK 293 cells are discussed in Section 6. Conclusions are given in Section 7.

2. Experimental Setups

The experimental study of generation of CAP jet is carried out in discharge devices with the coaxial and planar geometries. The plasma sources are shown in Figure 1. The coaxial device is a dielectric tube with a length of 100 mm and an inner diameter of 8 mm.

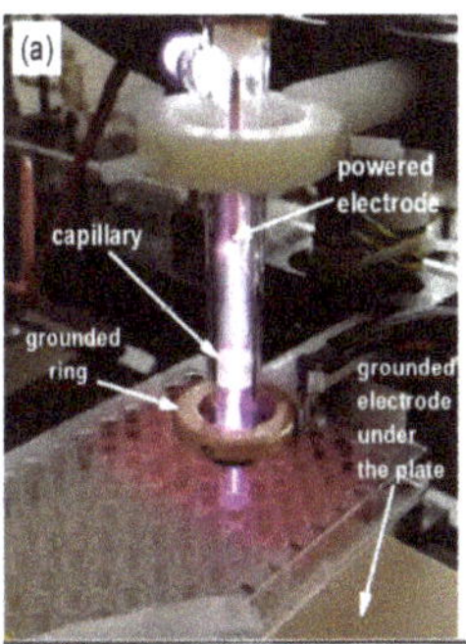

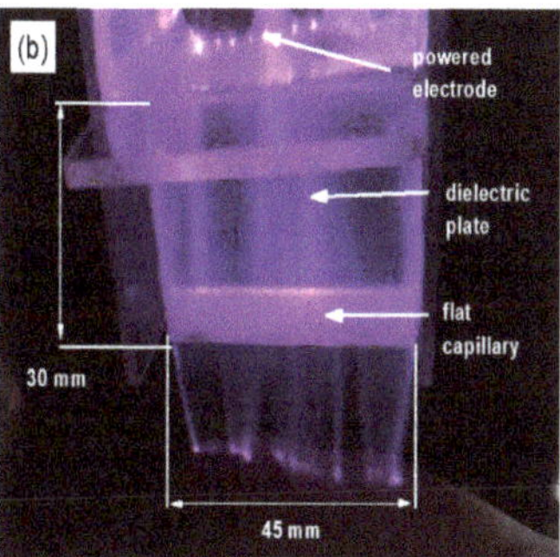

Figure 1. Plasma devices with the cylindrical (**a**) and planar (**b**) designs. The plasma jets are generated in helium.

A copper powered electrode with a length of 50 mm and with a diameter of 2 mm is inside of the dielectric tube. A capillary insert with a length of 6 mm and an inner diameter of 2.3 mm is placed some distance from the powered electrode. A copper grounded ring with an inner diameter of 18 mm is outside the quartz tube.

In the planar design of the plasma device, two quartz plates (2 mm × 30 mm × 45 mm) are inserted in the cylindrical dielectric frame. The gap between plates is 2 mm. The powered electrode is a copper multi-tip stripe located inside and the grounded electrode is outside. The capillary gap is 1 mm. As seen in Figure 1b, for the planar plasma device the treated area was elongated up to 45 mm.

In atmosphere, the helium plasma jet length is of 5–6 cm, but it can be elongated and redirected with a flexible dielectric tube. In Figure 2a,b, the images of plasma jets without and with a flexible tube inserted in the plasma device are shown. It is seen that the pathway of streamers can be controlled. The argon plasma jet shown in Figure 2c demonstrates an instability inside of the dielectric tube. The discharge current cords permanently change their trajectories inside the device tube in contrast to the helium quasi-stationary discharge glow (see Figure 2a).

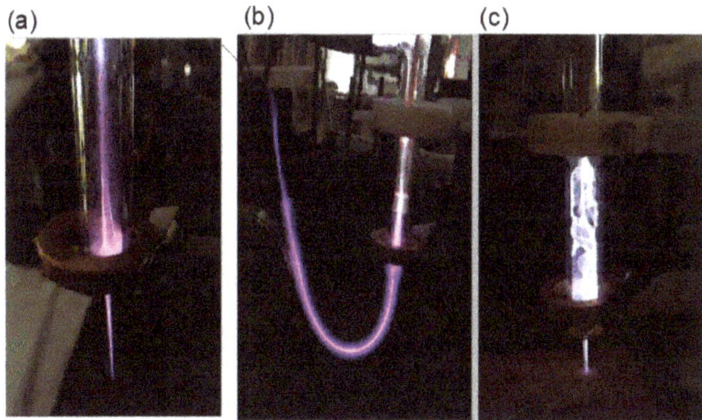

Figure 2. Images of cylindrical plasma devices and plasma jets generated in working gas helium (**a**), with a flexible tube (**b**) and in argon (**c**).

The gas system supplied working gases with the typical gas flow rate v of 1–10 L/min. A purity of helium and argon are of 99.995% and 99.998%, respectively. The power supply provided a sinusoidal voltage with a frequency of 40 kHz, voltage amplitude U up to 6 kV. In the experiment, the voltage and current of CAP were recorded with a Tektronix TDS 2024 oscilloscope with a passband of 200 MHz. The spectral composition of CAP was registered in the range of 200–750 nm, using a spectrometer with a resolution better than 0.5 nm.

3. Experimental Results on Plasma Jet

With increasing voltage amplitude, the discharge development in helium exhibits four steps. First, at a small voltage, luminous spots on a tip of the powered electrode appear. Then a glow spreads over the gap between the powered electrode and capillary insert. With further increasing U, the glow propagates inside the capillary and finally at some critical voltage U_{cr} the plasma jet starts to propagate beyond the dielectric tube over the inert gas flow in surrounding atmosphere. We found that in argon U_{cr} is essentially higher than in helium for the same gas flow rate v and diameter of capillary insert d. For example, for v = 2 L/min and d = 2.3 mm, U_{cr} is 2.6 kV for helium and U_{cr} = 3.5 kV for argon. Note that U_{cr} does not depend on the flow rate for $d > 2.3$ mm. The plasma jet length can be up to 60 mm in helium and only 20 mm in argon. An optimal geometry for the argon plasma jet is with the capillary placed at the edge of dielectric nozzle.

The experimental study of plasma jet generation with the planar one slit design source (see Figure 1b) shows the CAP formation picture which is similar to the coaxial geometry case. The feature of the planar design is that the presence of a multi-tip structure on the plane powered electrode is critical for the plasma jet formation. Please note that the critical voltage of CAP jet generation in helium for the planar source $U_{cr} \approx 6$ kV, which is much higher than U_{cr} for the cylindrical design.

The developed plane design of the CAP source allowed us to enlarge the zone exposed to the plasma jet. However, in our experiments with the cancer cell treatment discussed below we used the cylindrical plasma device to provide better uniformity of irradiation of wells with the media and cells without touching the plastic edge of wells.

In Figure 3, the waveforms of the voltage and current are shown for helium working gas. It is seen that the current increases at a positive half-wave of voltage as was observed in Ref. [4]. For all considered conditions of the CAP generation the discharge current does not exceed 10 mA.

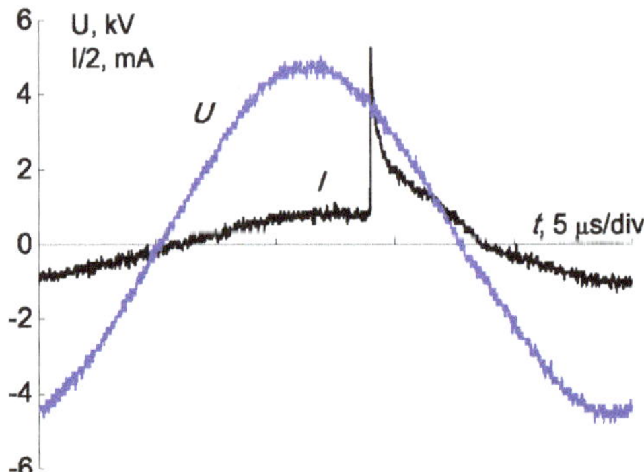

Figure 3. Discharge voltage and current with time in helium for $U = 5$ kV.

The examples of typical emission spectra of helium and argon CAP are shown in Figure 4. In addition to helium and argon lines in the plasma jet, there are lines of molecular nitrogen N_2, molecular nitrogen ions N_2^+, and nitric oxide NO. In the UV range, weak O_2 and O_2^+ lines are observed, as well as the Balmer series of the hydrogen line.

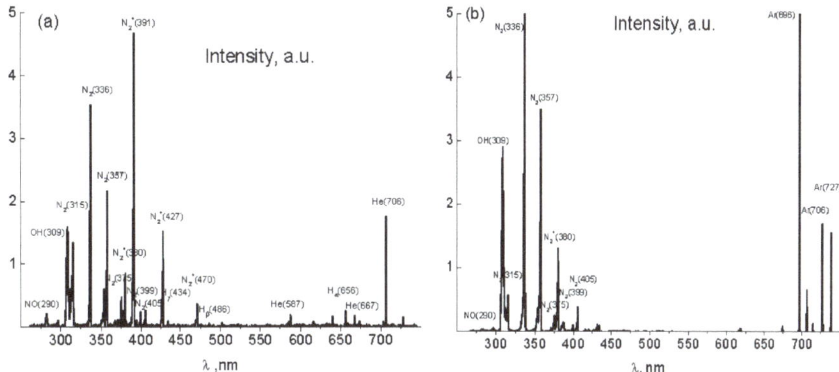

Figure 4. CAP spectra (**a**) in helium, v = 9.4 L/min, U = 5 kV, and (**b**) in argon, v = 3 L/min, U = 6.3 kV.

As seen in Figure 4, hydroxide OH appears in both spectra of helium and argon CAPs. In our experiments, the OH-radical production with CAP jet over the plasma-target interface was studied for the cases with and without the grounded substrate beneath the plate with the media and cells. As discussed below in Section 4, the presence of the grounded substrate must amplify the plasma-target interaction.

The spectra were measured in gas phase near the interface of plasma and liquid media. A variation of the intensity of OH-peak at λ = 309 nm in spectra was recorded for different voltages and gas flow rates. Please note that the peak at λ = 309 nm is always dominant in a group of OH-lines over our range of plasma parameters. A part of spectrum with OH-lines and the peak at 309 nm is shown in Figure 5 for the cases with and without the grounded substrate for U = 4.8 kV. It is seen that the intensity of OH-lines unexpectedly increases by an order of magnitude after the installation of the grounded substrate. An arrow in Figure 5 shows the OH-line at λ = 309 nm used for our analysis.

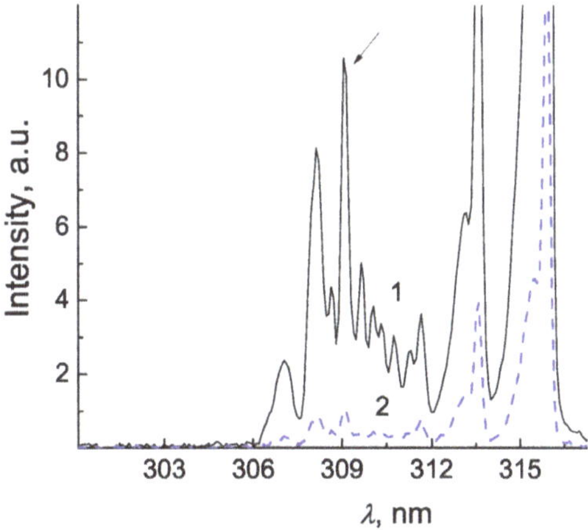

Figure 5. Part of spectrum with of OH peaks measured with the presence of the grounded substrate (1) and without it (2). An arrow indicates the peak at λ = 309 nm, helium, U = 4.9 kV, v = 3 L/min.

In the experiment, the spectra were measured within a range of the voltage from 2.5 kV to 6.5 kV. The intensity of the OH-peak depending on the voltage is given in Figure 6a. This intensity and consequently, the production rate of OH-radicals start to increase quickly for U > 4 kV. To illustrate the difference in the OH-production the ratio R of intensities of OH peaks with and without the grounded substrate is shown in Figure 6b. For a lower voltage, $U < 3.5$ kV, the occurrence of the substrate weakly affects the OH-radical production, whereas for $U > 3.5$ kV the ratio R exhibits a rise of a factor of 10.

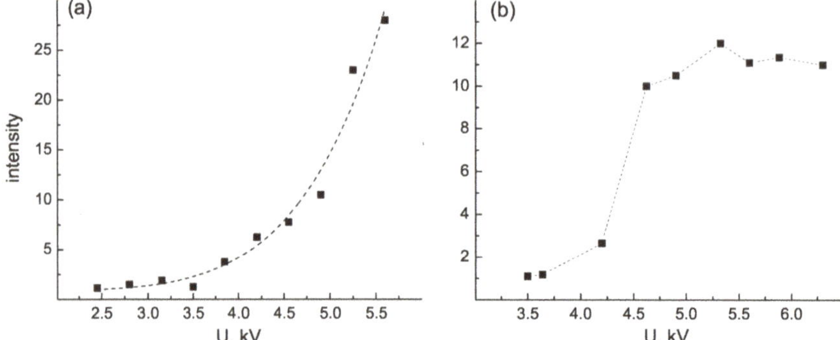

Figure 6. Intensity of OH-peak at $\lambda = 309$ nm in helium for different applied voltages with the grounded substrate (**a**) and the ratio of intensities of OH peaks with the grounded substrate and without it (**b**), v = 3 L/min.

The effect of gas flow rate on the intensity of OH-line at $\lambda = 309$ nm in spectrum is shown in Figure 7. The results of measurements indicate a range of flow rate from 2.5 L/min to 4 L/min is optimal for the OH-radical production.

In conclusion to this Section, the plasma devices for plasma jet generation with cylindrical and plane designs were developed and tested. The OH-radicals production rate was studied for different applied voltages and rates of gas flow. A change of the OH-radical intensity peak at $\lambda = 309$ nm was analyzed from the spectra recorded over the point of contact of the plasma jet and target. The comparison of the OH-peak in spectra measured (a) with the grounded substrate beneath the treated target and (b) for the electrically isolated target shows a considerable increase of OH-production for the case (a).

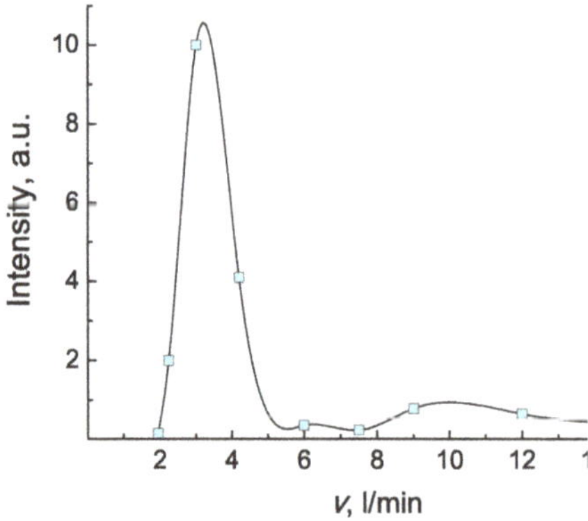

Figure 7. Intensity of OH-peak at $\lambda = 309$ nm in helium for different gas flow rates with the grounded substrate, $U = 4.9$ kV.

4. Effect of Grounded Electrode Simulation Results

The purpose of this theoretical study is to check the idea to use the grounded metal substrate under the bio-target to enhance the streamer characteristics near the target. The simulations of streamer formation in the plasma device and propagation to the target are performed in the framework of 2D fluid approach with the cylindrical symmetry. The fluid model of the discharge and streamer dynamics includes continuity equations for electrons, ions and mean electron energy and the Poisson equation for the electric potential and electric field. The transport coefficients in the continuity equations for electrons and their mean energy are calculated from the electron energy distribution function, which is calculated with solving the Boltzmann equation for a given E/N, where E is the strength of the electric field and N is the gas density, and a gas mixture. The detail model description is presented in Ref. [13]. A part of cylindrical simulation domain with the electric potential distribution is shown in Figure 8a. In simulations, as in the experiment, the streamer propagates to the target and interacts with it. The gas discharge ignition, streamer dynamics and interaction with the target were considered and compared for two positions of the (GR) substrate. In the first case, the GR substrate is placed 1 mm beneath the target and in the second case, the GR substrate is 500 mm from the target. The first and second cases refer to the experimental conditions with and without GR substrate, respectively.

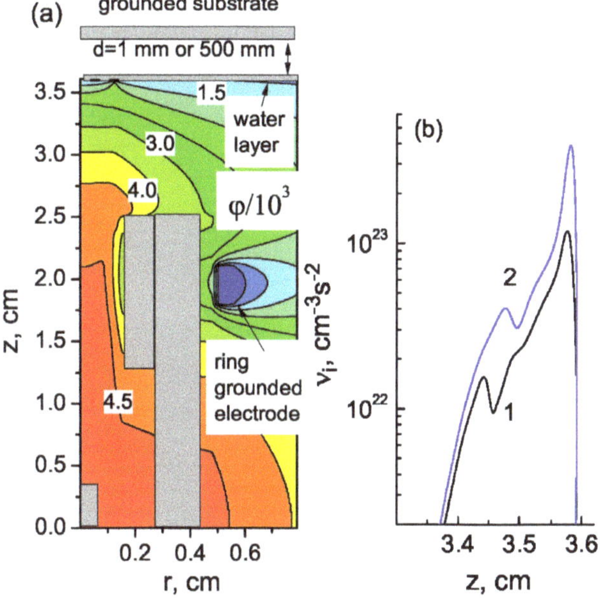

Figure 8. (a) Spatial distributions of electric potential at the moment when the streamer touches the target. Color palette is for 0–5000 V with the 500 V increment. (b) Ionization rate profile over z for the cases without the GR substrate (1) and with it (2), argon, $U = 4.9$ kV.

In simulation domain in Figure 8a, the target is at $z = 36$ mm. A depth of liquid layer is 5 mm. The radii of nozzle and dielectric tube are 1.15 mm and 2.4 mm, respectively. The length of dielectric tube is 25 mm. The dielectric permittivities of the quartz tube and liquid are 10 and 80, respectively. The radius of calculation domain is 60 mm. The powered electrode is embedded in the dielectric tube as in the experimental plasma device. The external ring grounded electrode with h = 3 mm and r = 4.8 mm is placed at $z = 17.5$ mm on the dielectric tube. The voltage amplitude U_0 is 4.9 kV and working gas is argon.

In simulations, the gradual voltage increase initiates the streamer formation near the tip of the powered electrode. The streamer propagates first inside of dielectric tube and then over the argon gas flow in the direction to the target. With approaching to the target, the streamer begins to accelerate and the ionization rate increases. The spatial potential distribution for the case without the GR substrate is shown in Figure 8a for the moment when the streamer touches the target. The potential gradually increases behind the streamer head in the direction of the powered electrode inside the dielectric tube. The position of the grounded ring electrode is seen as a potential dip. The streamer has a toroidal shape, and the point of streamer contact with the water layer is seen in Figure 8a as the potential bump with coordinates z = 3.58 cm and r = 1.25 mm. At this point the ionization rate has a maximum value. For this moment, the profiles of the ionization rate over z at r = 1.25 mm is shown in Figure 8b. In simulations, we found that the presence of the grounded substrate beneath the target essentially affects the streamer characteristics. It is seen in Figure 8b that the ionization rate is four times higher with the GR substrate (curve 1) compared to the case without it (curve 2). The explanation of an increase of the ionization rate can be found from the analysis of the potential distribution.

Figure 9 shows a zoom of the electric potential distribution near the liquid layer for two considered cases. The potential on the plasma-liquid interface $\varphi_w \approx 1$ kV without the GR substrate beneath and $\varphi_w = 372$ V with the GR substrate. It is seen that without the GR substrate beneath the target (when the grounded substrate is 500 mm apart from the target) the electric potential on the plasma-liquid interface ϕ_w is more elevated since the potential drop between the liquid layer and the remote GR substrate is relatively large. The potential drop over the liquid layer is negligible since $\epsilon_w = 80$. The electric field strength is determined with the potential gradient according the Poisson equation for the electric potential φ and electric field $\mathbf{E}$:

$$\mathrm{div}\nabla\varphi = 4\pi e(n_e - n_i), \mathbf{E} = -\nabla\varphi, \tag{1}$$

where n_e, n_i are electron and ion densities, e is the elementary charge. Thus, the elevation of φ_w on the plasma-water interface decreases the electric field near the surface and consequently the electron energy and the ionization rate.

The white line in Figure 9a indicates a trace of the maximum of the ionization rate during streamer propagation from the discharge tube to the target.

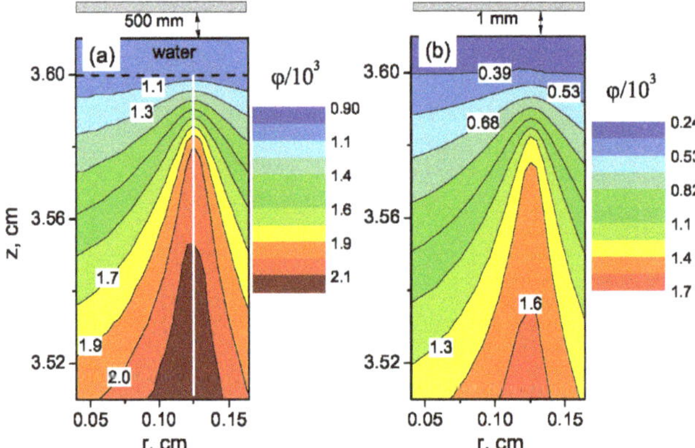

Figure 9. Spatial distributions of electric potential (**a**) without the GR substrate, color palette is for 900–2100 V, and (**b**) with the GR substrate, color palette is for 240–1700 V, argon, U = 4.9 kV.

The potential profiles (over white line in Figure 9a) for different times during the streamer approaching the surface is shown in Figure 10 for the cases with and without the GR substrate. A higher electrical field near the liquid surface for the case with the GR substrate is provided with a steeper potential drop compared to the case without the GR substrate.

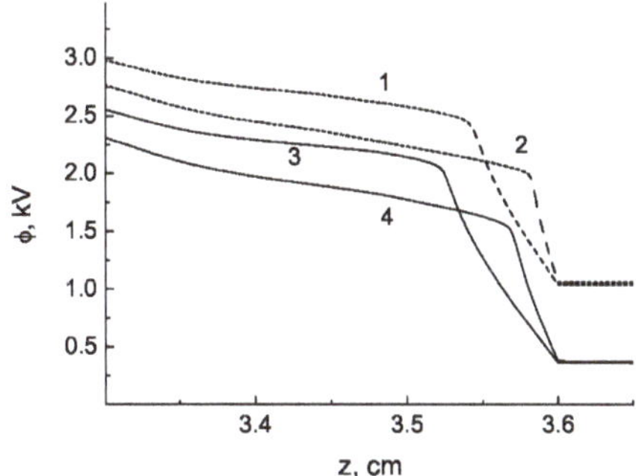

Figure 10. Potential profile over z for different times, t = 381.1 ns (1) and 382.8 ns (2) for the case without the grounded substrate and t = 374.5 ns (3) and 375 ns (4) for the case with GR substrate, argon, *U* = 4.9 kV.

The spatial distributions of the electron density, electron energy, normal component of electric field and ionization rate are shown in Figure 11 for enhanced plasma jet parameters with the GR substrate.

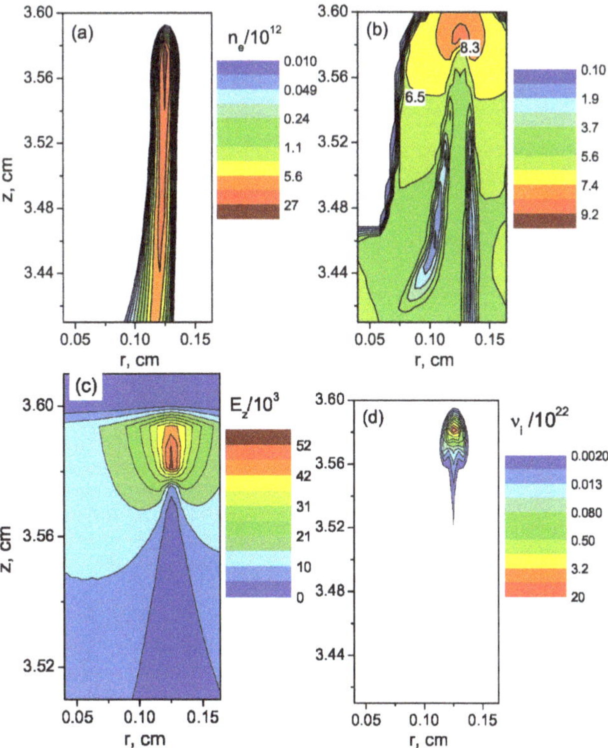

Figure 11. Spatial distributions of electron density, color palette is for $(10^{10}$–$2.7 \times 10^{13})$ cm^{-3}, log scale (**a**), electron energy, color palette, (0.1–9.2) eV (**b**), normal component of electric field E_z, color palette, $(5.2 \times 10^3$–$5.2 \times 10^4)$, V/cm (**c**) and ionization rate ν_i, color palette, $(10^{19}$–$2 \times 10^{23})$ cm^{-3}s^{-1}, log scale (**d**), with the GR substrate.

The maximum of the electron density and energy are 2.5×10^{13}, cm^{-3} and 8.3 eV. The E_z and E_r are 56 kV/cm and 26 kV/cm, respectively. If the GR substrate is absent these values are smaller that leads to decrease of the rate of plasma enhanced chemical reactions.

In conclusion of this section, in 2D fluid model simulations, the discharge ignition and streamer dynamics were studied for our experimental conditions. The streamer interaction with the target was calculated for the cases with and without grounded substrate beneath the target. It was shown that the electron density and energy, electric field strength and ionization rate considerably rise with the presence of the grounded substrate without increasing the discharge input power. This recommendation for the plasma jet enhancement was used in the cell-based experiments.

5. Materials and Methods for Study of Cells Response on CAP Treatment

Human cell lines, HEK 293, A431 and A549 were obtained from the American Type Culture Collection (ATCC; Manassas, VA, USA). HEK 293 (kidney embryonic cells) were grown in DMEM medium (Sigma-Aldrich, St. Louis, MO, USA), A431 (skin carcinoma) and A549 (human lung adenocarcinoma cells) were grown in DMEM F12 medium (Sigma-Aldrich, St. Louis, MO, USA) supplemented with 10% fetal bovine serum (FBS; Gibco BRL Co., Gaithersburg, MD, USA), 2 mM L-glutamine, 250 µg/mL amphotericin B, and 100 U/mL penicillin/streptomycin—complete medium. When cells were grown for the indirect treatment, all cell lines were cultured in DMEM F12 to unify the experimental conditions.

MTT Assay (Colorimetric assay, 3-(4,5-Dimethylthiazol-2-yl)-2, 5-diphenyltetrazolium bromide) was used to assess cytotoxicity and cell viability.

iCELLigence Assay. Cell proliferation and survival are monitored in real time with the iCELLigence RTCA (Real-Time Cell Analyser) system (ASEA Biosciences Inc., San Diego, CA, USA). The cells were seeded in 8-well E-plates with the integrated microelectronic sensor arrays and cell-to-electrode response is measured. The presence of adherent cells at the electrode-solution interface impedes electron flow. The magnitude of this impedance depends on the number, size and shape of cells. First, the cells were seeded at a density of 30,000 cells per a well in a total volume of 500 µL of DMEM F12. For the cytotoxicity assay, after the initial 24 h of cell proliferation, E-plates were pulled from iCELLigence RTCA system and the cells in an 8-well E-plates were treated with CAP jet. Since the diameter of wells on E-plate is larger than the diameter of the plasma jet the E-plates were periodically moving to ensure uniformity of treatment: E-plates were moved 3 times (rectangle-type moving): from the start point of irradiation plate was moved right for 5 mm, next down for 3 mm, next 6 mm left. Period (in seconds) of irradiation in each point was 0.25 of total irradiation. Then, E-plates were returned into the iCELLigence device and cell proliferation was monitored in real time for at least 100 h. Cell index, which reflects cell proliferation in real-time mode was calculated for each E-plate well by RTCA Software 1.2 (Roche Diagnosis, Meylan, France).

6. Viability of A549, A431 and HEK 293 Cells after CAP Jet Treatment

In the indirect treatment of cells with the irradiated medium the various sensitivity of different types of cells was observed. The human cell lines, A431 and HEK293 were seeded in 96-well flat-bottom plates at a density of 5000–10,000 cells in 100 µL of DMEM F12 media per well. The same cell-free DMEM F12 media in 12-well plates was exposed to the argon CAP jet during 5 min, 10 min and 15 min at $U = 4.9$ kV and v = 4 L/min. The indirect treatment of cells with the irradiated medium was carried out according to the scheme: 1:1, 50% of the initial medium and 50% of the medium treated with plasma. Therefore, 100 µL of treated media was added to each of 96-well flat-bottom plates with cells 15 min after the plasma treatment. To analyze the effect, MMT–Assay of Cell Viability was applied after the treatment. The viability of A431 cells and HEK 293 cells shown in Figure 12 was quantified 24 h after the media was added. A greater than 90% reduction in A431 cell viability was registered for the case of indirect treatment with the media exposure to the argon CAP jet during 5 min (see Figure 12a). For the cases of 10 min and 15 min, A431 cells viability is almost the same. In contrast to A431 cells viability, the HEK 293 cells shown in Figure 12b were less sensitive to the indirect CAP treatment. Indeed, the data showed 80% live cells after the treatment with the media exposed to CAP jet during 5 min. For the cases of 10 min and 15 min, the viability of HEK 293 cells decreases to 40% and 38%, respectively.

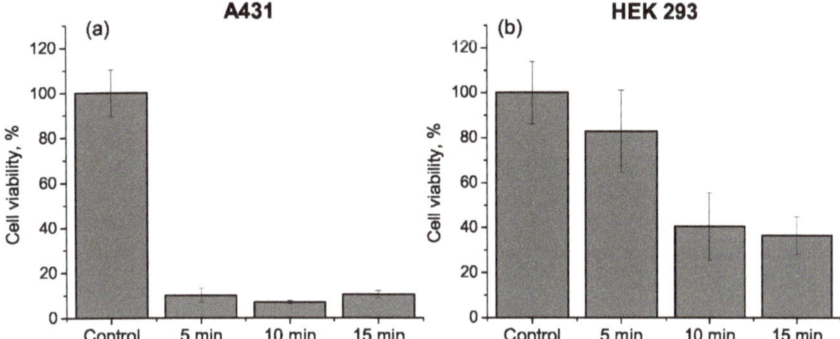

Figure 12. Live cells in % 24 h after the indirect argon CAP treatment of A431 human skin carcinoma (**a**) and HEK 293 human embryonic cells (**b**). Irradiation of media was 5, 10 and 15 min with CAP, without the grounded substrate, at U = 4.9 kV and v = 4 L/min.

The direct treatment with the CAP jet was carried out with A549 cells. In the experiment, A549 human lung adenocarcinoma cells were exposed to argon and helium CAP jets for different voltages, working gas flow rates and with and without the presence of the grounded substrate beneath the plates with the media and cells. MTT assay results in Figure 13 show the viability of A549 cells 24 and 48 h after the direct argon CAP treatment with and without the grounded substrate. The cells were exposed to the plasma jet during 1 min at U = 3.6, 4.2 and 4.9 kV.

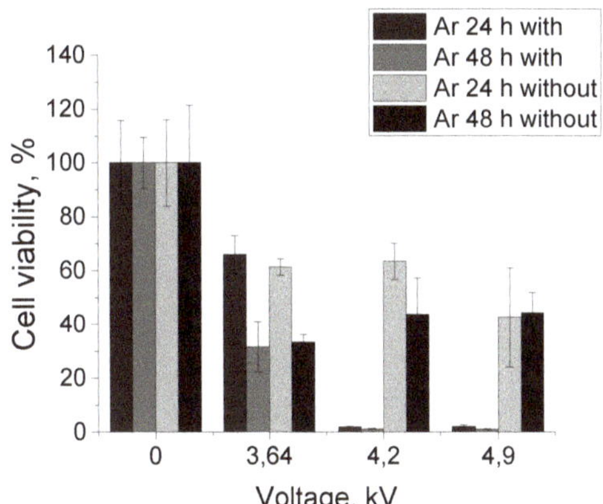

Figure 13. Number of survivals of A549 cells in % for cases with and without the grounded substrate after 24 and 48 h of direct argon CAP treatment, 1 min, at U = 3.6, 4.2 and 4.9 kV, gas flow rate v = 4 L/min.

For a lower voltage plasma treatment, at U = 3.6 kV, 60% cells survive 24 h after the treatment and approximately 35% after 48 h for both cases. So, no visible effect of the grounded substrate was registered. However, at higher voltages, at U = 4.2 kV and 4.9 kV, the use of the grounded substrate significantly enhances the effect of plasma treatment. In this case, the number of survivals drops to almost zero, whereas without the grounded substrate the statistics of cell viability remains the same for all voltages U = 3.6, 4.2 and 4.9 kV. It is seen that for the case of an electrically isolated target an increase of the input power in the plasma device does not enhance an efficiency of the plasma treatment.

The cell viability was also monitored with iCELLigence instrument to quantify cell proliferation. 30,000 cells were planted per a well of an iCELLigence plate the day before the experiment. A549 cells in an 8-well E-plates were treated for 2 min with argon and helium CAP jets. Taking different parameters of CAP, we wanted to show that the treatment does not work if the plasma parameters are chosen without optimization. First we made experiment with helium plasma jet parameters which are far from the optimal values found in Section 3, U = 4.2 kV and v = 5 L/min. For argon plasma jet the parameters of the plasma source U = 3.6 kV and v = 3 L/min. The cell-based experiments were carried out (a) with the grounded substrate under the plate with cells and (b) without a substrate. In Figure 14, the real-time generated outputs from the iCELLigence system before and after direct CAP jets treatment are shown. Cell index was calculated for each E-plate well by RTCA Software 1.2 (Roche Diagnosis, Meylan, France). A dip in Figure 14 refers to the moment when the plate was taken from iCELLigence instrument, irradiated by CAP jet and returned for analysis. Planted cells grow during 24 h before the treatment and 76 h after treatment. iCELLigence data show typical cell index curves that reflect cell proliferation in real-time mode.

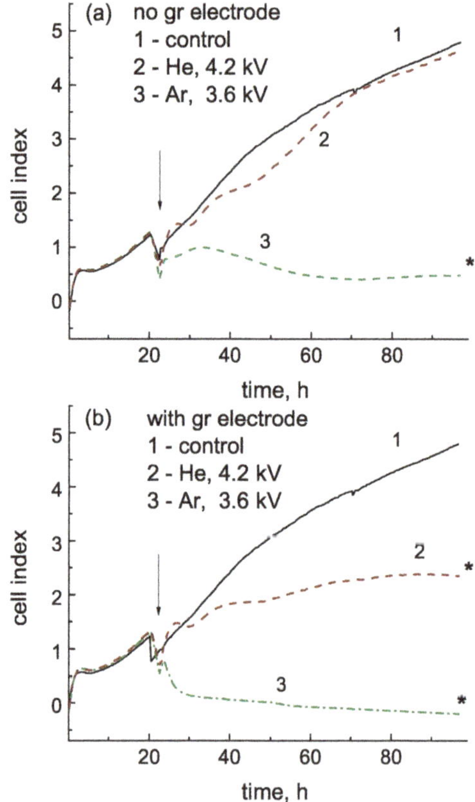

Figure 14. iCELLigence data showing typical Cell Index curves (CI) that reflect cell proliferation in real-time mode. Data are presented as average of three independent repeats. A549 cells were irradiated without the grounded substrate (**a**) or with it (**b**) for U = 4.2 kV and v = 5 L/min (helium) or 3.6 kV and v = 3 L/min (argon). The point at which cells were irradiated by CAP is indicated by the arrow. Students t-test was used to compare treatment effects, and the difference between control and groups was statistically significant at * $p < 0.05$.

For the case without the grounded substrate shown in Figure 14a, the viability of the cells exposed to helium plasma jet is practically the same as the viability of control cells, whereas for the argon plasma jet the number of survivals essentially diminishes. The low efficiency of helium CAP in this case can be explained by a large helium flow rate, v = 5 L/min. The presence of the grounded substrate enhances the effect of plasma treatment (see Figure 14b). For this case, the proliferation rate of the cells lowers to 52 % after the treatment with the helium CAP jet and the argon plasma jet treatment kills almost all cells after 24 h.

We found that the argon CAP has a greater impact on the cell viability at lower voltages compared to the helium CAP jet. As seen in Figures 13 and 14a, even without grounded substrate, the number of survivals quickly decreases during 48 h. An impact of the argon CAP jet at U = 3.6 kV looks less aggressive for the case without the grounded substrate, since the cell response begins after several hours after treatment, but not immediately, as with the grounded substrate. It means the U can be decreased for the former case.

To increase an efficiency of the helium CAP we use the optimal parameters for the OH-radical generation found in Section 3 and the grounded substrate beneath the plate with cells proposed in Section 4. According to experimental data shown in Figures 6b and 7, the OH-radical generation rate exhibits a maximum at U > 4.8 kV and v = 3 L/min. The optimal parameters were applied in the experiment with A549 cells. The plasma jet was generated at U = 5.5 kV, v = 3 L/min and with the grounded substrate beneath the plates. In Figure 15, the time-dependent viability measured with the iCELLigence system is shown before and after the helium CAP treatment.

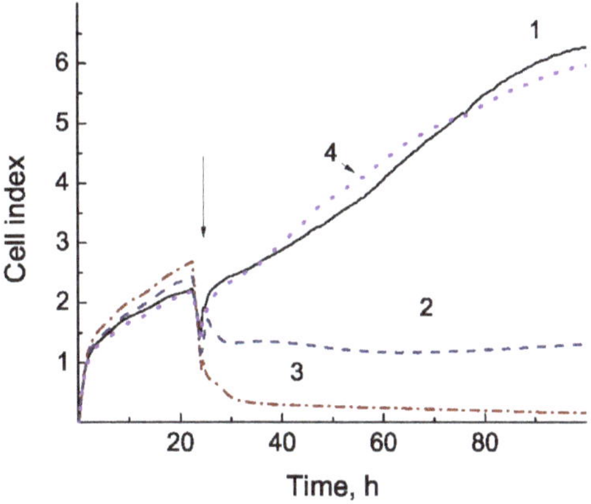

Figure 15. Cell Index of A549 cells before and after helium CAP treatment, with the grounded substrate for U = 5.5 kV and v = 3 L/min. 1—control, 2—direct (2 min), 3—direct (3 min), 4—indirect treatment with the media exposed to CAP during 2 min with the grounded substrate.

It is seen that after the 2 min and 3 min treatments at the same plasma parameters less than 25% and 2.4% of cells, respectively, survived during 100 h of monitoring. In Figure 15, the data of the indirect CAP treatment is also shown. The media was exposed to the same helium CAP jet during 2 min (with the grounded electrode) and added 15 min after to the wells with cells. The cell index for the case of this indirect treatment practically coincides with the control one. Thus, for indirect treatment the irradiation of media for 2 min, even with the presence of grounded substrate, is insufficient to affect cells.

In conclusion of this Section, the indirect CAP jet treatment of A431 and HEK293 cells with the irradiated medium (>5 min) showed various sensitivity of different types of cells. In the direct treatment, the A549 cancer cells were exposed to argon and helium CAP jets at different voltages and gas flow rates. The plate with the media and cells was placed on a) the dielectric electrically isolated plate or b) the metal grounded substrate. The viability of cells was monitored with MTT assay 24 and

48 hours after the CAP treatment and with iCELLigence Assay 100 h after the treatment. The results of measurements show the high efficiency of CAP treatment during 1–2 min with the presence of the grounded substrate and for optimal values of the voltage and gas flow rate. Note that there is no effect on cell proliferation rate after the indirect treatment with the media with 2 min irradiation in the case of grounded substrate.

7. Conclusions

The typical cold atmospheric plasma jet device with 40 kHz-driven AC voltage of 2.5–6.5 kV amplitude and with the cylindrical and plane designs were developed and tested. The inert gas (helium or argon) was pumping through the plasma device with the rate of 1–5 L/min. The cylindrical plasma device was used for the treatment of cancer cells in vitro.

In this work, for the first time, the grounded substrate was installed beneath the plate with the cells and media for the intensification of the plasma jet treatment. Previously, a strong dependence of the plasma jet characteristics near the surface of targets on their permittivities was reported (see, for example, Refs. [7,8]). The metal grounded target was shown can essentially intensify the plasma jet characteristics near the surface. In the experiments and 2D fluid model simulations, the interaction of CAP jet with (a) the bio-target placed on the grounded substrate and (b) electrically isolated bio-target without the metal substrate have been studied and compared. The presence of the grounded substrate was shown to lead to the considerable increase of the electric field over the plasma-media interface. More elevated electron energy and density and OH-radical production rate were registered with the grounded substrate compared to the case without it.

A range of optimal voltage and gas flow rate values for the plasma device for the efficient OH generation was obtained from the measurements of plasma spectra over the plasma-target interface. The intensities of OH-lines at $\lambda = 309$ nm were compared for different plasma jet characteristics and with and without the grounded substrate. When the plate with the media was placed on the grounded substrate, the OH-peak began to increase quickly at $U > 4$ kV and with further voltage it became 10 times higher than the OH-peak without the substrate. So, the voltage amplitude from 4.5 kV to 5 kV is optimal for the purpose of the OH-production. Please note that for a lower voltage, $U < 3.5$ kV, the occurrence of the substrate weakly affects the OH-peak in spectra. The measured effect of gas flow rate on the intensity of OH-line at $\lambda = 309$ nm showed that the gas flow rate ranging from 2.5 L/min to 4 L/min is optimal for the OH-radical production.

A drastic effect of the occurrence of the grounded substrate beneath the plate with cancer cells and media have been observed in the measurement of viability of cancer cells after the indirect and direct CAP treatment. The time-dependent viability of human cell lines, A431 (skin carcinoma), HEK 293 (kidney embryonic cells) and A549 (human lung adenocarcinoma cells) was monitored with MTT and iCELLigence assays after the plasma treatment with different CAP jet parameters. It was shown that the direct cancer cell treatment with CAP jets with the optimal plasma parameters and the grounded substrate visibly increased the impact of CAP.

Author Contributions: Conceptualization, I.S.; Methodology, D.Z., I.S. and O.K.; Software, I.S.; Validation and Investigation, D.Z., O.K., I.S., P.G., E.G., O.T. and E.Y.; Writing-Original Draft Preparation, I.S.

Funding: The authors gratefully acknowledge financial support from Russian Science Foundation, grant N 19-19-00255. The authors, EE and Dm Z, were partly supported financially by RFBR N 18-08-00510 for development of various designs of experimental setup.

Conflicts of Interest: The authors declare no conflict of interest.

References

1. Keidar, M.; Yan, D.; Sherman, J.H. *Cold Plasma Cancer Therapy*; Morgan&Claypool Publishers: San Rafael, CA, USA, 2019; p. 98.
2. Fridman, G.; Gutsol, F.A.; Friedman, G.; Shekhter, A.B. Applied plasma medicine. *Plasma Process. Polym.* **2008**, *5*, 503. [CrossRef]

3. Shashurin, A.; Shneider, M.N.; Dogariu, A.; Miles, R.B.; Keidar, M. Temporal behavior of cold atmospheric plasma jet. *Appl. Phys. Lett.* **2009**, *94*, 231504. [CrossRef]

4. Shashurin, A.; Keidar, M. Experimental approaches for studying non-equilibrium atmospheric plasma jets. *Phys. Plasmas* **2015**, *22*, 122002. [CrossRef]

5. Graves, D.B. Reactive species from cold atmospheric plasma: Implications for cancer therapy Plasma Process. *Plasma Process. Polym.* **2014**, *11*, 1120–1127. [CrossRef]

6. Babaeva, N.Y.; Kushner, M.J. Interaction of multiple atmospheric-pressure microplasma jets in small arrays: He/O_2 into humid air. *Plasma Sources Sci. Technol.* **2014**, *23*, 015007. [CrossRef]

7. Norberg, S.A.; Johnsen, E.; Kushner, M.J. Helium atmospheric pressure plasma jets touching dielectric and metal surfaces. *J. Appl. Phys.* **2015**, *118*, 013301. [CrossRef]

8. Akishev, Y.S.; Karalnik, V.B.; Medvedev, M.A.; Petryakov, A.V.; Trushkin, N.I.; Shafikov, A.G. How ionization waves (plasma bullets) in helium plasma jet interact with a dielectric and metallic substrate. *J. Phys. Conf. Ser.* **2017**, *927*, 012040. [CrossRef]

9. Sobota, A; Guaitella, O.; Sretenovi, G.B.; Kovaevi, V.V.; Slikboer, E.; Krsti, I.B.; Obradovi, B.M.; Kuraica, M.M. Plasma-surface interaction: Dielectric and metallic targets and their influence on the electric field profile in a kHz AC-driven He plasma jet. *Plasma Sources Sci. Technol.* **2019**, *28*, 045003. [CrossRef]

10. Ris, D.; Dilecce, G.; Robert, E.; Ambrico, P.F.; Dozias, S.; Pouvesle, J.-M. LIF and fast imaging plasma jet characterization relevant for NTP biomedical applications. *J. Phys. D: Appl. Phys.* **2014**, *47*, 275401.

11. Yue, Y.; Pei, X.; Lu, X. OH density optimization in atmosphericpressure plasma jet by using multiple ring electrodes. *J. Appl. Phys.* **2016**, *119*, 033301. [CrossRef]

12. Xu, G.; Liu, J.; Yao, C.; Chen, S.; Lin, F.; Li, P.; Shi, X.; Zhang, G.-J. Effects of atmospheric pressure plasma jet with floating electrode on murine melanoma and fibroblast cells. *Phys. Plasmas* **2017**, *24*, 083504. [CrossRef]

13. Schweigert, I.V.; Vagapov, S.; Lin, L.; Keidar, M. Enhancement of atmospheric plasma jet-target interaction with an external ring electrode. *J. Phys. D Appl. Phys.* **2019**, *52*, 295201. [CrossRef]

Article

The Role of Thermal Effects in Plasma Medical Applications: Biological and Calorimetric Analysis

Luigi Cordaro [1,2,*], Gianluca De Masi [1,3], Alessandro Fassina [1], Clarice Gareri [3], Antonio Pimazzoni [1,4], Daniele Desideri [5], Ciro Indolfi [3] and Emilio Martines [1]

[1] Consorzio RFX (CNR, ENEA, INFN, Università di Padova, Acciaierie Venete spa), Corso Stati Uniti 4, 35127 Padova, Italy; gianluca.demasi@igi.cnr.it (G.D.M.); alessandro.fassina@igi.cnr.it (A.F.); antonio.pimazzoni@igi.cnr.it (A.P.); emilio.martines@igi.cnr.it (E.M.)
[2] Centro Ricerche Fusione, University of Padova, Corso Stati Uniti 4, 35127 Padova, Italy;
[3] Division of Cardiology, Department of Medical and Surgical Sciences, Magna Graecia University of Catanzaro, Viale Europa, 88100 Catanzaro, Italy; clarice88@hotmail.it (C.G.); indolfi@unicz.it (C.I.)
[4] INFN-LNL, Viale dell'Università 2, 35020 Legnaro, Italy
[5] Department of Industrial Engineering, University of Padova, Via Gradenigo 6/a, 35131 Padova, Italy; daniele.desideri@unipd.it
*   Correspondence: luigi.cordaro@igi.cnr.it

Received: 29 October 2019; Accepted: 11 December 2019; Published: 17 December 2019

Abstract: Plasma Medicine tools exploit the therapeutic effects of the exposure of living matter to plasma produced at atmospheric pressure. Since these plasmas are usually characterized by a non-thermal equilibrium (highly energetic electrons, low temperature ions), thermal effects on the substrate are usually considered negligible. Conversely, reactive oxygen and nitrogen species (RONS), UV radiation and metastables are thought to play a major role. In this contribution, we compare the presence of thermal effects in different operational regimes (corresponding to different power levels) of the Plasma Coagulation Controller (PCC), a plasma source specifically designed for accelerating blood coagulation. In particular, we analyze the application of PCC on human blood samples (in vitro) and male Wistar rats tissues (in vivo). Histological analysis points out, for the highest applied power regime, the onset of detrimental thermal effects such as red cell lysis in blood samples and tissues damages in in-vivo experiments. Calorimetric bench tests performed on metallic targets show that the current coupled by the plasma on the substrate induces most of measured thermal loads through a resistive coupling. Furthermore, the distance between the PCC nozzle and the target is found to strongly affect the total power.

Keywords: atmospheric pressure plasma jet (APPJ); cold atmospheric plasma (CAP); plasma medicine; blood coagulation; tissue damage

---

1. Introduction

The basic idea of plasma medicine is to create a therapeutic effect based on a chemical rather than a thermal interaction with the living substrate. To do that, plasma medicine tools, through different schemes [1,2], produce a cold plasma in which the ion branch is kept at room temperature and the temperature increase of the target due to the plasma action is rather small. Such a cold plasma produces, interacting with the air, reactive nitrogen and oxygen species (RONS) which, together with electrons, electric field, UV radiation and metastable species, are considered the main factors responsible for the therapeutic action on biological tissues [3–5]. The recognized applications, most studied so far in the field of plasma medicine, are the disinfection of bacteria and fungi and decontamination [4–10], dermatological diseases [11–14], surface treatment of materials and bio-materials [15–18], wounds healing [10,13,19–21], blood coagulation [22–24] and the selective killing of cancer cells [13,25–27].

In the Dielectric Barrier Discharge (DBD) scheme, the one used in the Plasma Coagulation Controller (PCC) [23,28], plasma is usually produced by applying high voltage pulses at kHz frequencies between two close electrodes, separated by a dielectric layer. This last feature, together with the short duration (from hundreds ns to 1 μs) of the applied pulses, limits the amount of current flowing along the plasma to the target (and, of course, prevents the formation of an arc discharge). Indeed, reported measured currents on the target are usually of the order of few mA averaged over the entire duty cycle in a way that thermal effects are widely considered negligible.

However, even though a tentative standardization of the parameters within which plasma sources for bio-medical applications should operate has been proposed [29,30], a certain degree of variability is still present in the plasma medicine arena in terms of applied power during the treatment and in its relative effect on the living substrate.

Although the thermal effects induced by biomedical cold plasmas are typically considered negligible, we believe they should be quantified and properly taken into consideration, to be sure of working under such conditions that the plasma has no unwanted consequences. The aim of this work is to focus on the main mechanisms that determine such thermal effects and whether they can be deleterious for application to living matter. In our case, the flexibility of PCC enables a fine tuning of the power coupled to the plasma, making possible an extensive characterization of the thermal component. As it will be shown throughout this paper, this feature makes PCC a suitable plasma source for clinical applications.

This contribution has two main goals—from one, side we analyze the effect of PCC on biological samples highlighting the presence of thermal effects in specific operational conditions (Section 2); on the other side, we estimate the power delivered by PCC through different bench tests on metallic targets (Section 3). The basic idea behind this analysis is, thus, to clarify in the different PCC operational regimes whether thermal effects take place in the plasma action and which roles they could have.

2. The Effects of PCC on Biological Samples

The Plasma Coagulation Controller (PCC), an innovative plasma medicine tool, has been designed to promote non-thermal blood coagulation in patients where normal coagulation cannot act properly or sufficiently fast.

In the PCC plasma source, rapid periodic high voltage pulses are applied to a main electrode covered by a pyrex capillary closed at the end. A gas (typically helium or argon), flowing through a nozzle, is ionized by the induced electric field and then expelled, resulting in a plasma plume which can be directly applied to a substrate.

The main features of PCC has been described elsewhere [28] as well as the preliminary results on blood samples [23].

In this analysis, we study the effect of the PCC treatment at different powers on two biological samples—Male Wistar Rats tissues (in vivo experiments, Section 2.1) and blood samples taken from patients following anti-coagulant therapy (in vitro experiments, Section 2.2). This is possible thanks to the high flexibility of PCC allowing the modulation of the power coupled on the substrate to be treated—applied voltage ΔV and discharge repetition rate f can be easily varied through a graphic interface. For the sake of simplicity, in the following we compare three different operational regimes called, according to the applied power, LOW ($\Delta V = 6$ kV, $f = 5$ kHz), STD—standard ($\Delta V = 8$ kV, $f = 5$ kHz) and HIGH ($\Delta V = 8$ kV, $f = 10$ kHz). In the following we will refer to these acronyms. It must be said that, under none of the mentioned conditions, excessive heat is felt when the plasma touches the fingers.

2.1. In Vivo Experiments

2.1.1. Methods

Animal procedures have been performed conforming to the directive 2010/63/EU of the European Parliament and approved by the Italian Ministry of Health and by Institutional Animal Care and Use Committee of Magna Graecia University of Catanzaro, Italy. Male Wistar Rats, weighing 300–350 g, were randomly divided into four experimental groups—control (CTRL) that received no treatment, Low intensity (LOW), Standard intensity (STD) and High intensity (HIGH)—according to the PCC applied power. Rats from each experimental group have been anesthetized by intra-peritoneal injection of Zoletil (zolazepam hydrochloride and tiletamine hydrochloride; 20 mg/kg body weight) and Xylazine (10 mg/kg body weight). A slight cut was made at the femoral muscle level and the skin layer was removed. Right after, the PCC nozzle exit has been placed at 1cm of distance from the muscle, and a single application at the relative intensity has been performed. The rats were euthanized by an overdose of Zoletil (100 mg/kg body weight) and Xylazine (10%). Biopsies of femoral muscle of each experimental group have been withdrawn within 10′ and embedded in paraffin; 5 μm cross sections were prepared. Hematologist/Rosin (H/E) stainings were performed following the standard protocol.

2.1.2. Results

A rat model has been used to test the PCC potential side effects. After 90″ of PCC treatment directly on the femoral muscle a biopsy has been taken and processed as described in the method section. Representative images are reported in Figure 1.

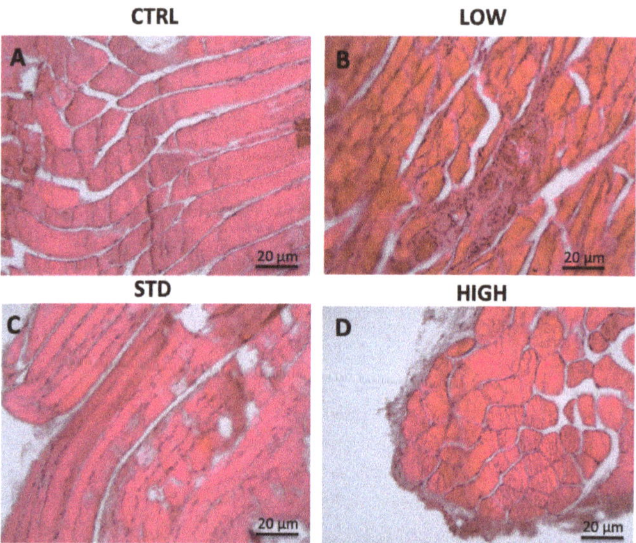

Figure 1. Plasma Coagulation Controller (PCC) effect on living tissue: representative images of H/E staining on rat muscle after PCC treatment. (**A**) CTRL (no treatment). (**B**) 90″ Low intensity treatment. (**C**) 90″ STD intensity treatment. (**D**) 90″ High intensity treatment.

As shown in Figure 1B, 90″ of low intensity treatment does not affect the muscle fibers, which look mainly like the control (Figure 1A). On the contrary, few nematodes formations appear following 90″ treatment at STD condition (Figure 1C); even if no major signs of damage, such as necrosis, are present. High intensity treatment has also been tested; in this case after 90″ of treatment disruption of collagen fibers and dehydration of muscle fibers were beginning, suggesting that this action could be too aggressive for a potential use in the clinical practice (Figure 1D).

2.2. In Vitro Experiments

2.2.1. Methods

Blood samples have been withdrawn, after obtaining informed consent, from patients undergoing anti-coagulant therapy, enrolled in an on-going study registry at Magna Graecia University. A 50 μL blood drop has been placed on the top part of a glass slide, placed at 1cm from the tip of the PCC, treated for different time points following distinctive conditions (LOW, STD, HIGH) and a smear obtained. The sample was left to dry in the air for about 1 h, fixed in formalin o/n and Hematoxylin/Eosin (H/E) stainings have been performed following standard protocol.

2.2.2. Results

The histological analysis performed on blood smear after PCC treatment has showed both platelets aggregation and fibrin polymerization, indicating a pro-coagulant effect of the plasma source on the blood, even ex-vivo, with all the setting tested. Indeed, 60″ at low intensity were sufficient to induce platelet aggregation and fibrin polymerization (Figure 2B) and the percentage of aggregation detected was proportional to the treatment intensity (Figure 2C,D). Despite the remarkable result observed for the blood coagulation, some concerning results appeared with the increase of intensity; after 60″ at STD conditions, red cell lysis begins (Figure 2C) and rapidly increase with longer treatments (data not shown). Also, 60″ of high intensity treatment cause a dramatic platelet aggregation and fibrin polymerization but also a robust red cell lysis. Of note, following longer treatments (more than 90″) the samples volume was strongly reduced, suggesting that, as expected, the heat developed has the tendency to dry the sample and possibly play a role in the coagulation process. Moreover, high intensity treatment also display peculiar black colored spots at the platelet clot areas, possibly due to the accumulation of iron released by the red cells.

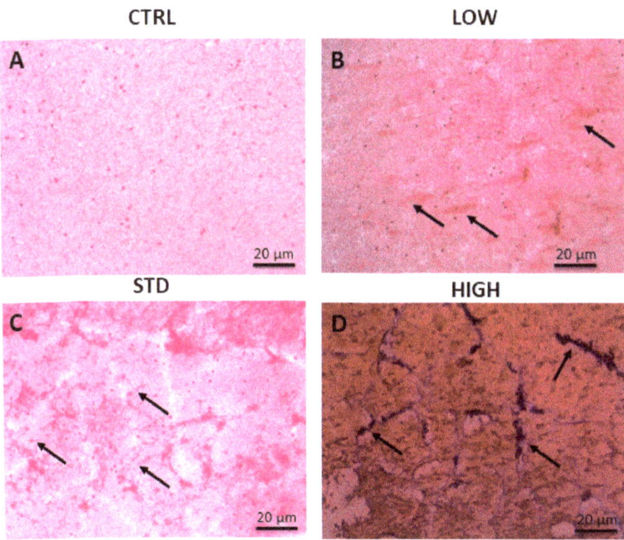

Figure 2. Blood Coagulation Analysis in vitro. Representative images of H/E staining on blood smears. (**A**) CTRL (no treatment). (**B**) 60″ LOW intensity treatment. (**C**) 60″ STD intensity treatment. (**D**) 60″ HIGH intensity treatment. The arrows indicate some platelet aggregates and fibrin networks.

2.3. Discussion

One of the first steps in healing process is the hemostasis, which prevents the continuous bleeding through the clot formation. In this work, we demonstrated how PCC treatments can induce platelet aggregation and fibrin formation, suggesting that this tool might be extremely useful for accelerating clot formation and subsequent wound healing. This is not surprising, since it is known that reactive oxygen and nitrogen species induce platelet aggregation [31,32]. Recently, it has been shown that low temperature plasma tools, like PCC, are able to promote the clot formation by activating different pathways, which include not only the platelets but also red cells aggregation as well as inflammatory process driven by white cells. In our case, longer treatments displayed red cell lysis [33,34], according to Ikehara et al. 2013 [33], while at shorter time points the hemolysis was absent, indicating that in vitro shorter treatments resulted more efficacious. At the same time, direct PCC usage on the skeletal muscle has not shown deleterious effects, at least for short treatments.

Taken together these data show a direct effect of the plasma source on blood samples, as well as on muscle tissues. In both cases, the effect observed was proportional to the treatment time and intensity, even if the extent of the consequence was different in the two targets. Indeed, at the same condition, the substrate of living tissue showed a lower effect compared to the blood samples. These results are very promising for a possible use of the PCC in the clinical routine and, on the other side, point out the appearance of thermal effects in some operational conditions linked to the treatment duration, the applied power on the target and the electric features of the target itself.

3. Calorimetric Measurements

In order to assess the heat flux which may be coupled by plasma to a surface placed in front of the PCC, calorimetric measurements have been performed by means of an infra-red (IR) camera. Metallic targets were used since their thermal and electrical properties are known. In order to identify the key mechanism for the power deposition, targets of different materials and size were considered. In addition, the dependence of the delivered power from the target distance was investigated. Targets were either left floating or connected to ground.

3.1. Experimental Setup

A picture of the experimental setup is given in Figure 3. The PCC is horizontally oriented and the plasma plume is fed to the target, a thin metal foil placed vertically. The target distance from the PCC can be modified by means of a micrometric translation stage. The IR camera looks at the metal target from the opposite side with respect to the PCC head, hence the plasma is not seen directly.

Figure 3. Experimental setup for calorimetric measurements. (**Left**) Plasma source and camera are placed on the opposite sides of the target. (**Right**) PCC plasma plume impinging on target. The distance between plasma source and substrate can be varied by means of a micrometer linear actuator.

Dealing with a small average power (few Watts), thin foils (from 40 to 80 µm) were used, so to have a sufficiently small heat capacitance. In fact, considering the power level of the PCC, the temperature increase of thick sample would be quite small and thus affected by considerable errors. The adopted metal sheets, made of steel and brass, had dimensions of $12.7 \times 12.7\,\mathrm{mm}^2$ and $12.7 \times 5\,\mathrm{mm}^2$. To increase the thermal emissivity up to $\epsilon = 0.8$, the surface observed by the IR camera was coated with a blackening layer (Molykote). The contribution of such a layer to the heat capacitance of the sample is considered negligible.

The IR camera (a FLIR A655sc) has a maximum sampling frequency of $f_s = 50$ Hz and a field of view of $15° \times 11.25°$ with a 640×480 pixel^2 sensor. The thermal power P_{cal} is estimated from the average temperature increase ΔT_{avg} of the target at $t = 0.5$ s after the plasma ignition as $P_{cal} = C\Delta T_{avg}/t$, where C is the target heat capacitance. Such a short measurement interval was chosen to limit both the underestimation of P_{cal} due to the convective losses towards air and the error due to timing at $1/f_s\,t$, that is 4%.

Typical time traces of the target average temperature are shown in Figure 4.

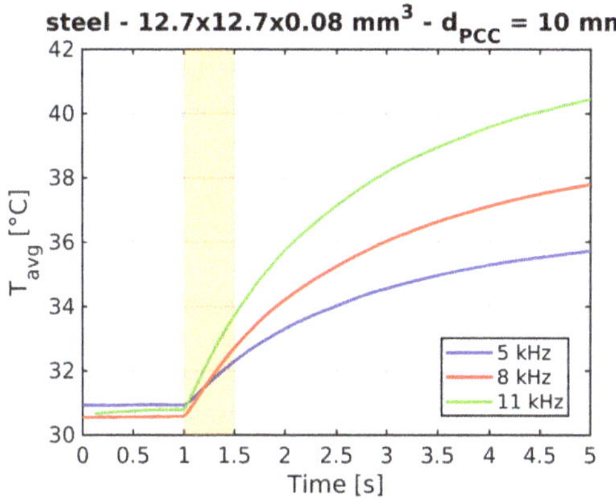

Figure 4. Time evolution of the average temperature of a stainless steel target ($12.7 \times 12.7 \times 0.08$ mm^3) at a distance $d_{PCC} = 10$ mm from the source nozzle, at different working frequencies. The transparency area indicates the time interval in which the thermal power P_{cal} is estimated.

3.2. Results

As a first step, the expected dependence of P_{cal} from the operational frequency was verified (see Figure 5a) and a linear fit was found to be suitable. The calculated slope A_{GR} for the grounded sample was found to be much larger (up to one order of magnitude) than the slope A_{FL} calculated for the floating sample. This result suggests that the key mechanism for the power deposition to the sample is the current flow within the target itself. This hypothesis is confirmed by the dependence of P_{cal} from the target thickness. In particular it was found that thicker samples have a significantly lower power deposition (see Figure 5b). Furthermore, for the same sample geometry, a smaller power (about 20% less) was measured on brass samples than on steel samples (steel electrical resistivity is about 70% larger).

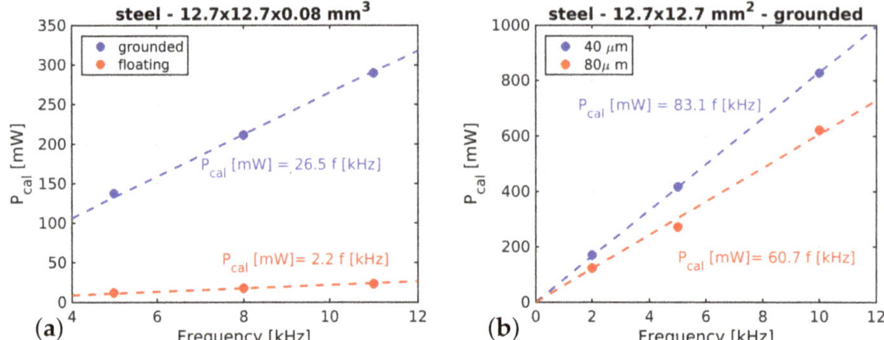

Figure 5. Calorimetric estimates of power at different frequency. (**a**) Comparison between grounded and floating samples. (**b**) Comparison between grounded samples of different thickness.

The dependence of P_{cal} on the target distance was found (for grounded samples) to be clear and monotonous as shown in Figure 6. In the first approximation, this is consistent with the lower interaction time between the plasma discharge (which propagates forward) and the substrate, as the distance source-substrate is increased. On the other hand, the plasma metal interaction region is strongly affected by the distance variation—this implies a difference in the resistive coupling, since the current patterns cannot be considered constant.

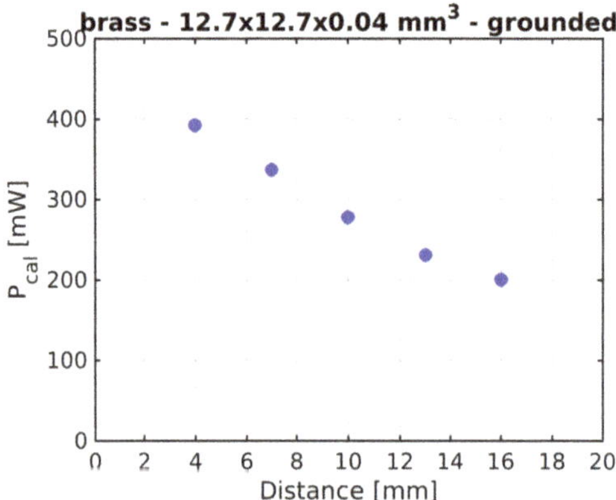

Figure 6. Calorimetric estimates of power at different distances from the PCC.

3.3. Discussion

The thermal power delivered by plasma has been determined, in several conditions of both power supply and distance, in different metallic foil samples. The performed analyses allowed to identify resistive coupling as the dominant mechanism for power deposition on the substrate. As a consequence, in order to precisely determine the heat load coupled upon a plasma treatment, it is necessary to infer the target resistivity. For this reason, extrapolating the load to biological samples is not trivial, since the substrate resistivity is not known and depends on the specific nature of the substrate. It could be useful, in this context, to use alternative materials, such as the soft gels described in Szili et al. (2014) [35], used as a model to simulate the transport of reactive species from plasma to biological tissues. However, even in that case, the thermal and electrical properties of such materials are almost unknown. So we preferred estimating the power carried by the plasma through the use of materials, of which we can precisely infer the chemical-physical properties.

It must also be said that the estimated power values are comparable with those reported in Weltmann et al. (2009) [36]. In that work, as in ours, it is shown how the power decreases with the distance from the source. In addition, the importance of controlling the plasma dose was emphasized to avoid deleterious effects on living tissues. To our knowledge, there are not many publications showing the thermal output of the plasma, like the one we reported. However, the importance of avoiding overheating of the biological substrate is widely recognized to avoid cellular damage [37,38]. Among the latter, we can report the denaturation of proteins, evaporation of cellular liquid, structural alterations of the cell membrane, up to cell death [37,39]. The heat dose control is therefore fundamental for the clinical applications of cold plasmas. This control can be done in multiple ways—by varying the power coupled to the plasma, controlling the distance and the treatment time. In this context, PCC is a useful source of plasma for biomedical applications, thanks to its flexibility. We have shown that it is indeed possible to vary the power coupled to the plasma in such a way as to induce coagulation but avoiding, at the same time, the appearance of deleterious effects on biological tissues. Our calorimetric analysis therefore serves as a preliminary activity for tuning such a plasma dose. It must be said however that a direct thermographic measurement on the biological sample could be helpful to avoid undesired thermal loads. Furthermore, it shall be noticed that hand-held use of plasma sources would hardly grant a mm precision in the distance source-substrate, while mechanical suspension has to be preferred where possible (e.g., in vitro).

4. Conclusions

The effectiveness of cold plasmas in inducing a wide range of biological and therapeutic effects is demonstrated by a large and ever growing body of scientific literature. In most cases, it is generally assumed that thermal effects are negligible. This assumption is mostly based on the evaluation of the power deposition to the tissues performed through calorimetric measurements on targets of known material, typically metals. Also the existing international standards prescribe an evaluation of power transfer using copper targets of known features.

The objective of the present work was to evaluate the effectiveness of the PCC plasma source at different power levels, in relation to the task of inducing non-thermal blood coagulation. Our in vitro and in vivo tests showed that the plasma, at low power and for short time treatments, is indeed able to effectively accelerate platelet aggregation and fibrin formation, thus inducing coagulation. However, at higher powers and for longer treatments, also harmful effects appear, such as red cells lysis, with destruction of collagen fibers and dehydration of muscle fibers. It is worth mentioning that, in our case, higher powers and longer treatments have been applied only for characterization purposes, while are not necessary for obtaining therapeutic effects. On the other hand, this study demonstrates the importance of controlling the plasma dose for clinical applications, in order to avoid the appearance of tissue damage.

In carrying out this task, it was ascertained that the main heating mechanism is due to the electric current flowing in the sample. This leads to the conclusion that a power deposition evaluation performed on targets of given material could not be representative of what happens when the plasma is applied to actual living tissues. Indeed, the effect is expected to be remarkably different than in the case of metal targets. One might be tempted to prescribe the use of target materials which more closely resemble actual tissues in terms of resistivity, however also in this case a large uncertainty may be expected, due to the large variations of resistivity among different biological samples, especially when one includes conditions such as the presence of blood or other fluids. Furthermore, grounding of the treated sample is an issue which also affects power deposition.

Overall, these findings point to the necessity of having plasma sources where input power can be varied, as is the case for the PCC, so that it can be adjusted to the substrate conditions. Furthermore, being the thermal load and not the dissipated power the relevant quantity to be adjusted, one may envision, for the next generation of devices, methods to locally measure the substrate temperature and feedback systems to adjust power levels and/or distance between source and substrate so as to keep this quantity to a desired value [40]. The identification of the best technical solutions for the development of this new generation of tools will be the task for plasma physicists and engineers working in the plasma medicine field for the years to come.

Author Contributions: Conceptualization, L.C., G.D.M., A.F., C.G., A.P. and E.M.; Data curation, L.C., A.F., C.G. and A.P.; Formal analysis, C.G. and A.P.; Funding acquisition, G.D.M. and C.I.; Investigation, L.C., G.D.M., A.F., C.G., A.P. and D.D.; Project administration, G.D.M., C.I. and E.M.; Resources, D.D., C.I. and E.M.; Software, A.P.; Supervision, C.I. and E.M.; Validation, L.C., G.D.M., A.F., C.G., A.P., D.D. and E.M.; Visualization, L.C., A.F., C.G. and A.P.; Writing original draft, L.C., G.D.M., A.F., C.G. and A.P.; Writing review & editing, L.C., G.D.M., A.F., C.G., A.P., D.D. and E.M.

Funding: This work was supported in part by the University of Padova—Department of Industrial Engineering, under Grant year 2017—BIRD171931 and in part by a research project named "A novel plasma medicine tool for accelerated haemostasis", which has received funding from the "Fondazione Con Il Sud" within the call "Brains2South".

Acknowledgments: The authors thank the Bionem Laboratory of the Department of Experimental and Clinical Medicine (University "Magna Graecia" of Catanzaro) for its technical support in the 3D printing activity of the PCC source; M. Maniero and M. Fincato for their valuable support during the experimental campaigns.

Conflicts of Interest: The authors declare no conflict of interest.

References

1. Weltmann, K.D.; Kindel, E.; von Woedtke, T.; Hähnel, M.; Stieber, M.; Brandenburg, R. Atmospheric-pressure plasma sources: Prospective tools for plasma medicine. *Pure Appl. Chem.* **2010**, *82*, 1223–1237. [CrossRef]

2. Von Woedtke, T.; Reuter, S.; Masur, K.; Weltmann, K.D. Plasmas for medicine. *Phys. Rep.* **2013**, *530*, 291–320. [CrossRef]

3. Laroussi, M. Plasma medicine: A brief introduction. *Plasma* **2018**, *1*, 47–60. [CrossRef]

4. Kong, M.G.; Kroesen, G.; Morfill, G.; Nosenko, T.; Shimizu, T.; Van Dijk, J.; Zimmermann, J. Plasma medicine: An introductory review. *New J. Phys.* **2009**, *11*, 115012. [CrossRef]

5. Martines, E. Interaction of cold atmospheric plasmas with cell membranes in plasma medicine studies. *Jpn. J. Appl. Phys.* **2019**, *59*, SA0803, doi:10.7567/1347-4065/ab4860. [CrossRef]

6. Brun, P.; Pellizzaro, A.; Cavazzana, R.; Cordaro, L.; Zuin, M.; Martines, E. Mechanisms of Wound Healing and Disinfection in a Plasma Source for the Treatment of Corneal Infections. *Plasma Med.* **2017**, *7*, 147–157. [CrossRef]

7. Martines, E.; Zuin, M.; Cavazzana, R.; Gazza, E.; Serianni, G.; Spagnolo, S.; Spolaore, M.; Leonardi, A.; Deligianni, V.; Brun, P.; et al. A novel plasma source for sterilization of living tissues. *New J. Phys.* **2009**, *11*, 115014, doi:10.1088/1367-2630/11/11/115014. [CrossRef]

8. O'Connor, N.; Cahill, O.; Daniels, S.; Galvin, S.; Humphreys, H. Cold atmospheric pressure plasma and decontamination. Can it contribute to preventing hospital-acquired infections? *J. Hosp. Infect.* **2014**, *88*, 59–65, doi:10.1016/j.jhin.2014.06.015. [CrossRef]

9. Hertwig, C.; Meneses, N.; Mathys, A. Cold atmospheric pressure plasma and low energy electron beam as alternative nonthermal decontamination technologies for dry food surfaces: A review. *Trends Food Sci. Technol.* **2018**, *77*, 131–142, doi:10.1016/j.tifs.2018.05.011. [CrossRef]

10. Brehmer, F.; Haenssle, H.; Daeschlein, G.; Ahmed, R.; Pfeiffer, S.; Görlitz, A.; Simon, D.; Schön, M.; Wandke, D.; Emmert, S. Alleviation of chronic venous leg ulcers with a hand-held dielectric barrier discharge plasma generator (PlasmaDerm® VU-2010): Results of a monocentric, two-armed, open, prospective, randomized and controlled trial (NCT 01415622). *J. Eur. Acad. Dermatol. Venereol.* **2015**, *29*, 148–155. [CrossRef]

11. Metelmann, H.-R.; Vu, T.T.; Do, H.T.; Le, T.N.B.; Hoang, T.H.A.; Phi, T.T.T.; Luong, T.M.L.; Nguyen, T.T.H.; Nguyen, T.H.M.; Nguyen, T.L.; et al. Scar formation of laser skin lesions after cold atmospheric pressure plasma (CAP) treatment: A clinical long term observation. *Clin. Plasma Med.* **2013**, *1*, 30–35. [CrossRef]

12. Emmert, S.; Brehmer, F.; Hänßle, H.; Helmke, A.; Mertens, N.; Ahmed, R.; Simon, D.; Wandke, D.; Maus-Friedrichs, W.; Däschlein, G.; et al. Atmospheric pressure plasma in dermatology: Ulcus treatment and much more. *Clin. Plasma Med.* **2013**, *1*, 24–29. [CrossRef]

13. Bernhardt, T.; Semmler, M.L.; Schäfer, M.; Bekeschus, S.; Emmert, S.; Boeckmann, L. Plasma Medicine: Applications of Cold Atmospheric Pressure Plasma in Dermatology. *Oxidative Med. Cell. Longev.* **2019**, *2019*, 3873928. [CrossRef] [PubMed]

14. Chutsirimongkol, C.; Boonyawan, D.; Polnikorn, N.; Techawatthanawisan, W.; Kundilokchai, T. Non-Thermal Plasma for Acne and Aesthetic Skin Improvement. *Plasma Med.* **2014**, *4*, 79–88. [CrossRef]

15. Stoffels, E.; Flikweert, A.J.; Stoffels, W.W.; Kroesen, G.M.W. Plasma needle: A non-destructive atmospheric plasma source for fine surface treatment of (bio)materials. *Plasma Sources Sci. Technol.* **2002**, *11*, 383–388, doi:10.1088/0963-0252/11/4/304. [CrossRef]

16. Malinowski, S.; Herbert, P.A.F.; Rogalski, J.; Jaroszyńska-Wolińska, J. Laccase Enzyme Polymerization by Soft Plasma Jet for Durable Bioactive Coatings. *Polymers* **2018**, *10*, 532, doi:10.3390/polym10050532. [CrossRef] [PubMed]

17. Malinowski, S.; Wardak, C.; Jaroszyńska-Wolińska, J.; Herbert, P.A.F.; Panek, R. Cold Plasma as an Innovative Construction Method of Voltammetric Biosensor Based on Laccase. *Sensors* **2018**, *18*, 4086, doi:10.3390/s18124086. [CrossRef]

18. Malinowski, S.; Jaroszyńska-Wolińska, J.; Herbert, P.A.F. Theoretical insight into plasma deposition of laccase bio-coating formation. *J. Mater. Sci.* **2019**, *54*, 10746–10763, doi:10.1007/s10853-019-03641-2. [CrossRef]

19. Fridman, G.; Friedman, G.; Gutsol, A.; Shekhter, A.B.; Vasilets, V.N.; Fridman, A. Applied Plasma Medicine. *Plasma Process. Polym.* **2008**, *5*, 503–533. [CrossRef]

20. Haertel, B.; Von Woedtke, T.; Weltmann, K.D.; Lindequist, U. Non-thermal atmospheric-pressure plasma possible application in wound healing. *Biomol. Ther.* **2014**, *22*, 477. [CrossRef]

21. Schmidt, A.; Bekeschus, S.; Wende, K.; Vollmar, B.; von Woedtke, T. A cold plasma jet accelerates wound healing in a murine model of full-thickness skin wounds. *Exp. Dermatol.* **2017**, *26*, 156–162. [CrossRef] [PubMed]

22. Ikehara, S.; Sakakita, H.; Ishikawa, K.; Akimoto, Y.; Yamaguchi, T.; Yamagishi, M.; Kim, J.; Ueda, M.; Ikeda, J.I.; Nakanishi, H.; et al. Plasma blood coagulation without involving the activation of platelets and coagulation factors. *Plasma Process. Polym.* **2015**, *12*, 1348–1353. [CrossRef]

23. De Masi, G.; Gareri, C.; Cordaro, L.; Fassina, A.; Brun, P.; Zaniol, B.; Cavazzana, R.; Martines, E.; Zuin, M.; Marinaro, G.; et al. Plasma Coagulation Controller: A Low- Power Atmospheric Plasma Source for Accelerated Blood Coagulation. *Plasma Med.* **2018**, *8*, 245–254. [CrossRef]

24. Fridman, G.; Peddinghaus, M.; Balasubramanian, M.; Ayan, H.; Fridman, A.; Gutsol, A.; Brooks, A. Blood Coagulation and Living Tissue Sterilization by Floating-Electrode Dielectric Barrier Discharge in Air. *Plasma Chem. Plasma Process.* **2006**, *26*, 425–442. [CrossRef]

25. Martines, E.; Brun, P.; Artico, R.; Cavazzana, R.; Cordaro, L.; Fischetto, D.; Zuin, A.; Zuin, M. Intracellular Rons Increase Induced By Helium Plasma Triggers Apoptosis In Cancer Cells. *Clin. Plasma Med.* **2018**, *9*, 28–29. [CrossRef]

26. Keidar, M.; Shashurin, A.; Volotskova, O.; Ann Stepp, M.; Srinivasan, P.; Sandler, A.; Trink, B. Cold atmospheric plasma in cancer therapy. *Phys. Plasmas* **2013**, *20*, 057101. [CrossRef]

27. Metelmann, H.R.; Seebauer, C.; Rutkowski, R.; Schuster, M.; Bekeschus, S.; Metelmann, P. Treating cancer with cold physical plasma: On the way to evidence-based medicine. *Contrib. Plasma Phys.* **2018**, *58*, 415–419. [CrossRef]

28. Cordaro, L.; De Masi, G.; Fassina, A.; Mancini, D.; Cavazzana, R.; Desideri, D.; Sonato, P.; Zuin, M.; Zaniol, B.; Martines, E. On the Electrical and Optical Features of the Plasma Coagulation Controller Low Temperature Atmospheric Plasma Jet. *Plasma* **2019**, *2*, 156–167. [CrossRef]

29. International Electrotechnical Commission. *IEC 60601-2-76:2018 Medical Electrical Equipment—Part 2-76: Particular Requirements for the Basic Safety and Essential Performance of Low Energy Ionized Gas Haemostasis Equipment*; International Electrotechnical Commission: Geneva, Switzerland, 2018.

30. Mann, M.S.; Tiede, R.; Gavenis, K.; Daeschlein, G.; Bussiahn, R.; Weltmann, K.D.; Emmert, S.; von Woedtke, T.; Ahmed, R. Introduction to DIN-specification 91315 based on the characterization of the plasma jet kINPen® MED. *Clin. Plasma Med.* **2016**, *4*, 35–45. [CrossRef]

31. Pignatelli, P.; Pulcinelli, F.M.; Lenti, L.; Gazzaniga, P.P.; Violi, F. Hydrogen peroxide is involved in collagen-induced platelet activation. *Blood* **1998**, *91*, 484–490. [CrossRef]

32. Bekeschus, S.; Schmidt, A.; Weltmann, K.D.; von Woedtke, T. The plasma jet kINPen—A powerful tool for wound healing. *Clin. Plasma Med.* **2016**, *4*, 19–28. [CrossRef]

33. Ikehara, Y.; Sakakita, H.; Shimizu, N.; Ikehara, S.; Nakanishi, H. Formation of membrane-like structures in clotted blood by mild plasma treatment during hemostasis. *J. Photopolym. Sci. Technol.* **2013**, *26*, 555–557. [CrossRef]

34. Miyamoto, K.; Ikehara, S.; Takei, H.; Akimoto, Y.; Sakakita, H.; Ishikawa, K.; Ueda, M.; Ikeda, J.I.; Yamagishi, M.; Kim, J.; et al. Red blood cell coagulation induced by low-temperature plasma treatment. *Arch. Biochem. Biophys.* **2016**, *605*, 95–101. [CrossRef] [PubMed]

35. Szili, E.J.; Bradley, J.W.; Short, R.D. A 'tissue model' to study the plasma delivery of reactive oxygen species. *J. Phys. Appl. Phys.* **2014**, *47*, 152002. [CrossRef]

36. Weltmann, K.D.; Kindel, E.; Brandenburg, R.; Meyer, C.; Bussiahn, R.; Wilke, C.; Von Woedtke, T. Atmospheric pressure plasma jet for medical therapy: Plasma parameters and risk estimation. *Contrib. Plasma Phys.* **2009**, *49*, 631–640. [CrossRef]

37. Zenker, M. Argon plasma coagulation. *GMS Krankenhaushygiene Interdiszip.* **2008**, *3*.

38. Keidar, M.; Walk, R.; Shashurin, A.; Srinivasan, P.; Sandler, A.; Dasgupta, S.; Ravi, R.; Guerrero-Preston, R.; Trink, B. Cold plasma selectivity and the possibility of a paradigm shift in cancer therapy. *Br. J. Cancer* **2011**, *105*, 1295. [CrossRef]

39. Kalghatgi, S.U.; Fridman, G.; Fridman, A.; Friedman, G.; Clyne, A.M. Non-thermal dielectric barrier discharge plasma treatment of endothelial cells. In Proceedings of the 2008 30th Annual International Conference of the IEEE Engineering in Medicine and Biology Society, Vancouver BC, Canada,20-25 Aug. 2008; pp. 3578–3581.

40. Gidon, D.; Graves, D.B.; Mesbah, A. Predictive control of 2D spatial thermal dose delivery in atmospheric pressure plasma jets. *Plasma Sources Sci. Technol.* **2019**, *28*, 085001. [CrossRef]

 *applied sciences*

Article

Direct Treatment of Liquids Using Low-Current Arc in Ambient Air for Biomedical Applications

Vladislav Gamaleev [1,*], Naoyuki Iwata [1], Masaru Hori [2], Mineo Hiramatsu [1] and Masafumi Ito [1,*]

1 Department of Electrical and Electronic Engineering, Meijo University, Nagoya 468-8502, Japan
2 Center for Low-Temperature Plasma Sciences, Nagoya University, Nagoya 468-8502, Japan
* Correspondence: vlad.gamaleev@gmail.com (V.G.); ito@meijo-u.ac.jp (M.I.);
  Tel.: +81-52-832-1151 (ext. 5085) (M.I.)

Received: 30 July 2019; Accepted: 22 August 2019; Published: 26 August 2019

Abstract: In this work, we developed a portable device with low production and operation costs for generating an ambient air low-current arc (AALCA) that is transferred to the surface of a treated liquid. It was possible to generate a stable discharge, irrespective of the conductivity of the treated liquid, as a sequence of corona, repeating spark, and low-current arc discharges. The estimated concentration of reactive oxygen and nitrogen species (RONS) in plasma-treated water (PTW) produced using AALCA treatment was two orders of magnitude higher than that of PTW produced using conventional He nonequilibrium atmospheric pressure plasma jets or dielectric barrier discharges. The strong bactericidal effect of the treatment using AALCA and the water treated using AALCA was confirmed by survival tests of *Escherichia coli*. Further, the possibility of treating a continuous flow of liquid using AALCA was demonstrated.

Keywords: plasma; RONS; low-current arc; water treatment; antimicrobial activity

---

1. Introduction

Recently, treatment of liquids using various atmospheric pressure plasmas has attracted much attention owing to its wide range of possible applications, such as synthesis of nanocarbons and nanoparticles, sterilization of microorganisms, growth enhancement of plants, plasma farming, cancer therapy, and water purification [1–13]. Typically, low-temperature nonequilibrium atmospheric pressure plasma jets (NEAPPJs) or dielectric barrier discharges (DBDs) are used for activation of liquids by plasma [5,12–16]. However, in both the case of atmospheric pressure plasma jet (APPJ) and DBD, the production rate of reactive oxygen and nitrogen species (RONS), which play key roles in bactericidal effect and plant growth promotion, is low and irradiation of the liquid takes a long time (from several minutes to hours) due to the low density of the plasma [5,12–21]. Moreover, in most cases, plasma is generated using pure gases (e.g., argon, helium, or oxygen) at a high voltage, so a substantial amount of electrical energy is required to sustain the discharge and only a limited volume (typically several milliliters) of medium can be treated at one time [4,7,17,19–22]. The low efficiency, the requirement of pure gases, and the limited volume of liquid that can be treated make plasma-treated water (PTW) expensive and its use in medicine and agriculture impossible. The most promising method for effective and low-cost production of PTW is treating liquids with high-density plasma generated by using air as a process gas [3,18,23–26]. Furthermore, it may be possible to scale up production and reduce the cost of PTW by applying plasmas generated in ambient air to treat a continuous flow of liquid. Several works have reported on treating a liquid flow using air as the process gas, however, the major drawback of these studies is the considerably low production rate of RONS, which limits the wide application of the produced PTW [13,15,18,27,28].

Typically, cold nonequilibrium plasmas are used for biomedical and agricultural applications to avoid thermal damage of the treated target [12,13,29,30]. However, in the case of using liquids treated by plasma, a plasma-treated medium (PTM) could be stored after treatment to reduce the temperature and for use after storage considering the presence of the long-living RONS [31,32]. For that reason, thermal and warm plasmas could be used for irradiation of medium following the cooling of PTM and use in biomedical applications. Considering the high efficiency of arc or gliding arc discharges in the decomposition of CO_2 and other gases and the effective production of RONS, this type of discharge could also be employed for the effective generation and fast delivery of RONS to the target liquid [3,25,33,34]. There are many works on the generation of arc and discharges and other types of plasmas inside or on the surface of highly conductive liquids using various configurations and experimental conditions, however, generation of high-density plasma transferred to the surface of deionized water (DI) or other liquids with low electrical conductivity in ambient air is still challenging owing to the relatively high electric currents required for the stable generation of the discharge [10,11,18,23,35–39]. It appears that it is possible to generate a low-current arc in ambient air that can be transferred to the surface of a liquid that has low electrical conductivity by using a high-voltage pin electrode as the anode and the target liquid as the cathode. In this work, we developed a simple and compact generator for direct treatment of liquids using an ambient air low-current arc (AALCA) and analyzed the discharge process and production rate of RONS.

2. Materials and Methods

2.1. Experimental Setup

The experimental setup is presented in Figure 1. A plasma discharge was generated in ambient air between the needle electrode (cut of stainless-steel wire, 751487, Nilaco, Tokyo, Japan) and the surface of the liquid by applying a positive high voltage to a needle electrode (anode). A grounded ring electrode (Ni wire, NI-311386, Nilaco, Tokyo, Japan) was introduced into the liquid, and during the discharge, the grounded liquid was used as a cathode. Using the needle electrode as the anode allowed us to avoid strong heating and melting of the electrode.

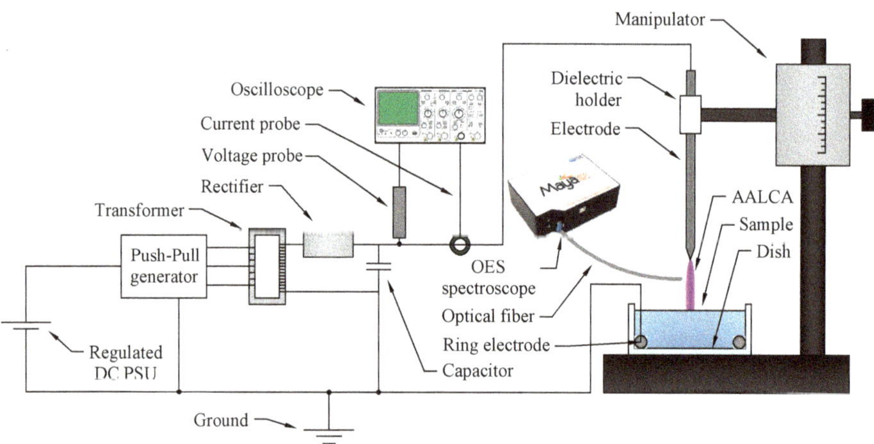

Figure 1. Experimental setup.

A high voltage was produced by a custom power supply consisting of a regulated direct current power supply (DC PSU), a push–pull generator, a high-voltage transformer, a diode rectifier, and a reservoir capacitor. The voltage and current waveforms during the discharge were monitored using a high-voltage probe (P6015A, Tektronix, Beaverton, OR, USA) and a current probe (2877, Pearson Electronics Inc., Palo Alto, CA, USA), and stored in a digital oscilloscope (TBS2000, Tektronix

Beaverton, OR, USA). The discharge gap between the surface of the liquid and the pin of the needle electrode was set at 8.5 mm using a micromanipulator.

The optical emission spectrum from the plasma was recorded using a multichannel spectrometer (Maya 2000 Pro, Ocean Optics, Largo, FL, USA) with a measurement range of 200–665 nm and a quartz optical fiber (QP1000-2-SR, Ocean Optics, Largo, FL, USA).

2.2. Investigation of Properties of the Treated Water

DI water (18.2 MΩ·cm at 25 °C) was produced by a Direct-Q 3UV (Millipore, Tokyo, Japan) system. To compare the performance of the developed AALCA to commonly used plasma sources, DI water was treated using AALCA, a commercially available NEAPPJ (PN-110/120TPG, NU Global Co., Ltd., Nagoya, Japan) [40], and a high-density nonequilibrium atmospheric pressure radical source (NEAPRS, Tough Plasma FPA10, Fuji MFG Co., Ltd., Tokyo, Japan) [4,31,41].

The concentration of H_2O_2 in the produced PTW was measured using high-performance liquid chromatography (HPLC) with an electrochemical detector (ECD-700s, Eicom, Kyoto, Japan) and a column (5020-07146 Inertsil CX, 5 µm, 4.6 × 250 mm, GL Sciences, Torrance, CA, USA). Concentrations were estimated from the measured HPLC spectra of PTW using the calibration curve obtained from H_2O_2 (081-04215, FUJIFILM Wako Pure Chemical Corporation, Osaka, Japan) diluted to various concentrations.

Concentrations of NO_2^- and NO_3^- were measured using ion chromatography with a conductivity detector (CDD-10Avp, Shimadzu, Kyoto, Japan) and a column (Shim-pac IC-A3, 4.6 × 150 mm, Shimadzu, Kyoto, Japan). Concentrations were estimated from data measured by the ion chromatography spectra of PTW using the calibration curve obtained from a reference solution (228-33603-93, Shimadzu, Kyoto, Japan) diluted to various concentrations. The composition of the reference solution is presented in Table 1.

Table 1. Composition of reference solution.

Component	Anion	Concentration (mg/L)
NaF	F^-	5
NaCl	Cl^-	10
$NaNO_2$	NO_2^-	15
KBr	Br^-	10
KNO_3	NO_3^-	30
KH_2PO_4	PO_4^{3-}	30
Na_2SO_4	SO_4^{2-}	40

The pH value of PTW was measured using a pH meter (S SevenCompactTM pH/Ion, METTLER TOLEDO, Columbus, DC USA) and a probe (InLab Pure Pro-ISM, METTLER TOLEDO, Columbus, DC, USA). The pH meter and probe were calibrated using a standard buffer solution set (101-S, Horiba, Kyoto, Japan). The conductivity of PTW was measured using a conductivity measurement cell (LAQUA 3551-10D, Horiba, Kyoto, Japan).

2.3. Microorganism Culturing and Colony Counting

A survival test of *Escherichia coli* was performed to evaluate the bactericidal efficacy of direct treatment using AALCA and the bactericidal effect of the produced PTW. *E. coli* (O1:K1:H7) was precultured in 3 mL of nutrient broth (DifcoTM, BD, San Jose, CA, USA) at 250 rpm with a regulated temperature of 30 °C for 17 h. Then, *E. coli* was separated by centrifugation at 5000 × *g* for 3 min and diluted with DI water to obtain a concentration of bacteria of approximately 1×10^7 mL^{-1}.

To investigate the bactericidal effect of treatment using AALCA, the prepared *E. coli* suspension was treated with AALCA under various conditions. To investigate the bactericidal effect of the produced PTW, a water sample was treated with AALCA and a 0.3 mL *E. coli* suspension was mixed with 2.7 mL of the produced PTW and incubated at 250 rpm with a regulated temperature of 30 °C for up to 24 h. The 0.3 mL *E. coli* suspension was mixed with 2.7 mL of pristine DI water as the control.

Then, the analyzed suspensions of *E. coli* were diluted, and samples of 0.1 mL were spread onto nutrient agar medium (DifcoTM, BD, San Jose, CA, USA) and cultured at 37 °C for 24 h. The number of *E. coli* that survived after the treatment was estimated using a colony-forming units (CFU) assay.

3. Results and Discussion

3.1. Generation of the Discharge

It was found that it was possible to generate various types of discharges between the pin electrode and the liquid depending on the input power. In the case of a low input power of 6 W supplied to the push–pull generator, it was possible to generate a stable positive corona. During the corona discharge, the current was negligible, resulting in a charge of the reservoir capacitor at the high-voltage side and stabilization of the voltage at about 11.25 kV, as shown in Figure 2a. A photo of the corona discharge is shown in Figure 2b. With an increase in input power to 6.5 W, the voltage during the corona discharge increased to about 12 kV, resulting in the generation of occasional spark discharges. The moment of breakdown and the number of consequent spark discharges were random. After breakdown, the voltage dropped to 1–2 kV and a short current pulse was observed during the spark discharge. After termination of the spark discharge, the current dropped to 0 following oscillation and the voltage increased to the initial value. Owing to a low input power, it took several tens of milliseconds to charge the reservoir capacitor and reach the initial voltage. A further increase of power resulted in an increase in the number of occasional spark discharges, and finally at an input power of 9.5 W, consequent spark discharges were observed, as shown in Figure 2c,d. In the case of the 9.5 W input power, the spark discharges were similar to those observed for the lower input powers. On the basis of the current and voltage (I and V) waveforms, the type of discharge was the same. Some variation of peak current and voltage and frequency observed in the I and V waveforms were related to the migration of the discharge spot on the surface of the liquid. In the case of repeating spark discharges, the sparks had a scanning area of about 5 mm in diameter on the surface of the liquid, as shown in Figure 2d.

A further increase of power up to 23 W did not change the discharge type and the same type of repeating sparks was observed, however, the increase of the input power resulted in some changes of the current and voltage waveforms, as shown in Figure 2e. An increase of input power resulted in an increase of peak current during the spark discharge, which decreased the time required for charging the reservoir capacitor, causing an increase of frequency of the spark discharges. On the other hand, the increase of input power resulted in stabilization of the spot of the spark and the repeating spark discharges looked like steady glow plasma (Figure 2f).

Increasing the input power to values above 23 W resulted in the occasional formation of low-current arc discharges, and a stable repulsing low-current arc discharge was generated at an input power of 25 W. Compared with the repeating spark discharges, the repulsing low-current arc discharge featured a lower peak voltage of about 2.5 kV, continuous current during the discharge, and high stability of the current and voltage waveforms, as shown in Figure 2g. A photo of AALCA at the input power of 25 W is presented in the Figure 2h. In the present setup, a low-pass filter was not employed, resulting in an AC ripple of voltage, which can be observed in Figure 2e,g.

In the case of the direct application of an input power of 25 W, the discharges occurred in the same order as described above, that is, as a sequence of a corona discharge, repeating sparks with increasing frequency, and finally, a low-current arc discharge. At the moment of voltage application, a corona discharge was generated first and was followed by repeating spark discharges for several seconds. Corona and spark discharges delivered RONS and ions originating from the air, evaporated liquid, and electrodes to the liquid, resulting in an increase of the electric conductivity of the liquid in a similar way to what was reported elsewhere [18,24,42]. With the increase of the conductivity of treated DI water, occasional pulsed arc discharges appeared and finally, after 17–20 s of treatment, the discharge changed from repeating sparks to the stable AALCA.

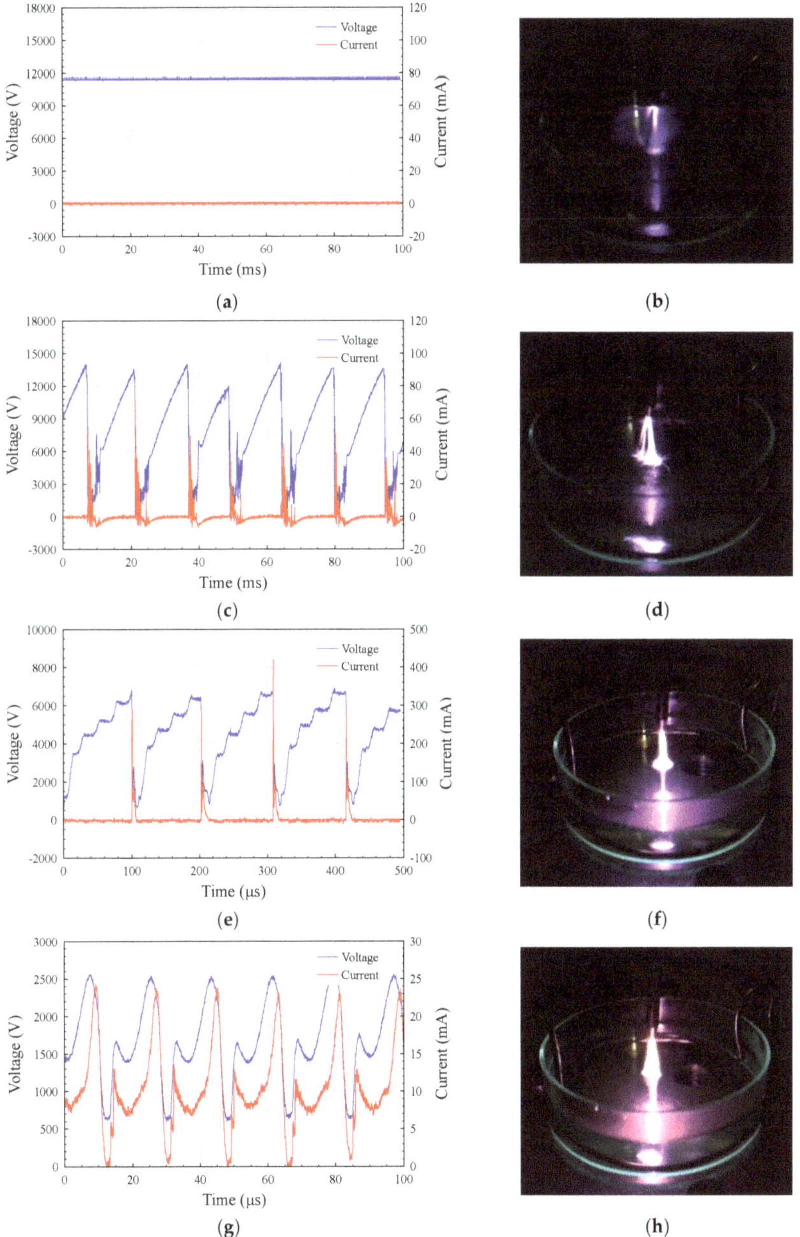

Figure 2. I & V waveforms (**a**) and photo (**b**) of corona discharge, I and V waveforms (**c**) and photo (**d**) of spark discharges at 9.5 W of input power, I and V waveforms (**e**) and photo (**f**) of spark discharges at 23 W of input power, and I & V waveforms (**g**) and photo (**h**) of ambient air low-current arc (AALCA) at 25 W of input power.

Figure 3 shows the conductivity of 5 mL of DI water as a function of the duration of the treatment by AALCA. The conductivity of the water sample increased from about 0.1 μS/cm for pristine DI water to 190 μS/cm after treatment for 20 s by spark discharges, which enabled the generation of a low-current arc that was transferred to the surface of the DI water, which initially had low conductivity.

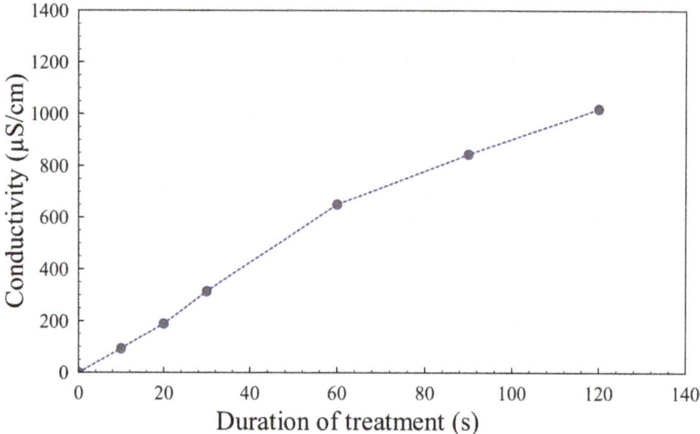

Figure 3. Conductivity of deionized water (DI) water as a function of the duration of plasma treatment.

Moreover, the pH value gradually decreased from 6.76 for pristine DI water to 2.89 after treatment for 2 min, as shown in Figure 4, which is a considerably low value as compared with the PTW reported elsewhere [18,24,31,43]. The low pH value suggests a high concentration of RONS and possibly some ions (delivered from the electrodes by erosion) in the treated liquid, which could explain the increase of conductivity after the plasma treatment.

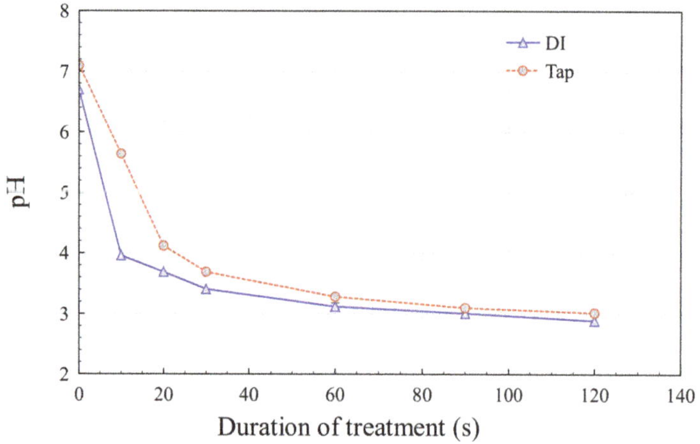

Figure 4. pH value of DI and tap water as a function of the duration of plasma treatment.

The presented approach allows generation of a stable AALCA discharge that is transferred to the surface of a liquid, even if it has initially low electrical conductivity (such as DI water), without preparation and modification of the liquid. Variation of conductivity affected only the duration of the state of the repeating spark discharges (17–19 s for DI water with an initial conductivity of about 0.1 μS/cm and 0–2 s for tap water with an initial conductivity in the range of 170–250 μS/cm).

The main drawback of using AALCA is the heating and evaporation of the treated liquid, which is used as the cathode and interacts with high-energy ions and other charged species. Thermographic observations showed that the temperature of the 5 mL DI water sample treated by AALCA did not exceed 45 °C (see Figure 5a) after 1 min of treatment and 70 °C (see Figure 5b) after 5 min of treatment.

The relatively low temperature of the small volume of liquid treated with AALCA could be explained by local evaporation of the liquid. During treatment, the liquid was heated to the boiling temperature locally and evaporated when a bulk liquid was heated at a lower rate and remained below the boiling temperature. The higher temperature of the liquid in the area of contact with the plasma can be observed from thermographic images (Figure 5a,b).

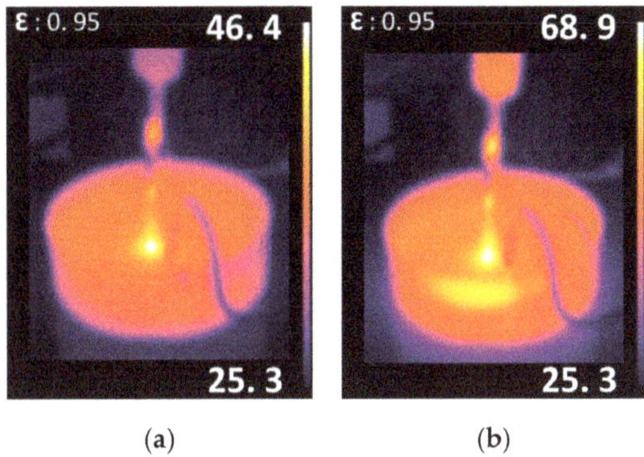

Figure 5. Thermographic image of DI water treated with AALCA for 1 min (**a**) and 5 (**b**) min.

It was confirmed that 1.2 mL of water was evaporated from the original 5 mL sample after 5 min of treatment. To ensure stable generation of AALCA even after partial evaporation of the liquid and change in the length of the discharge gap, 28 W of input power was used in all the experiments described below.

3.2. Species Generated in Gas Phase

Optical emission spectrometry (OES) in the wavelength region from 200 to 665 nm was used to investigate the main excited species generated by AALCA. OES spectra of AALCA is presented in Figure 6.

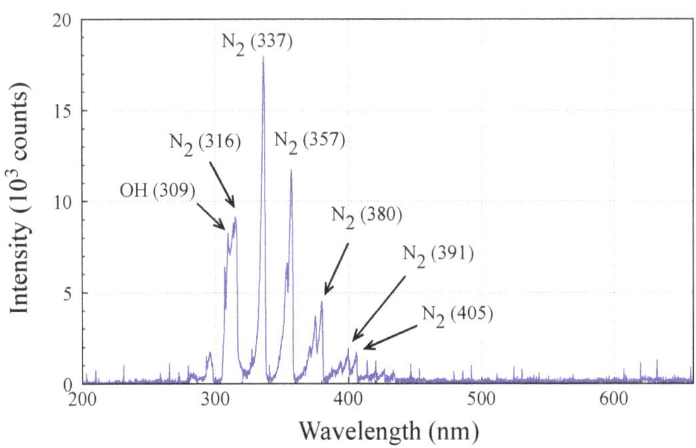

Figure 6. Optical emission spectrometry (OES) spectra of AALCA.

Owing to the high concentration of nitrogen in the air, the observed spectrum was dominated by N_2 (C→B) emissions as a result of many excitation processes in the plasma. As shown in Figure 6, the optical emissions from the N_2 second positive system ($C_3\Pi \rightarrow B_3\Pi$) at 316, 337, 357, 380, and 405 nm; the N_2 + first negative system ($B_2\Sigma_u + \rightarrow X_2\Sigma_g +$) at 391.4 nm; and the OH emissions from the transition of $A_2\Sigma+ \rightarrow X_2\Pi$ at 309 nm were clearly observed for AALCA [44]. Evaporation of the liquid discussed above could be the source of OH radicals, which could be produced by electron impact dissociation of water molecules [14]. Despite the presence of hydrogen from evaporated water, hydrogen emission lines (e.g., H_α and H_β) were not observed in the spectra.

The observed OES spectrum that was dominated by N_2 emissions was similar to emission spectra observed in other types of air plasmas transferred to the surface of water reported elsewhere [18,24]. The OES results indicated the presence of excited OH radicals and atomic nitrogen in the AALCA which could be delivered to the liquid from the gas phase and converted to other RONS in the solution, such as hydrogen peroxide (H_2O_2), nitrite (NO_2^-), and nitrate (NO_3^-).

3.3. Generation of RONS in Water

To check the production efficiency of RONS in the liquids treated with AALCA and compare it to commercially available plasma sources, samples of 5 mL DI water were treated with AALCA, NEAPPJ, and NEAPRS. The sample was treated with AALCA at an input power of 28 W for 30 and 60 s. NEAPPJ was operated using He gas with a flow rate of 3 slm and a distance from the nozzle of the plasma jet to the surface of the liquid of 20 mm, ensuring contact of plasma with liquid and the effective delivery of RONS. NEAPRS was operated with a mixture of Ar (4.75 slm), N_2 (200 sccm), and O_2 (50 sccm) at a total flow rate of 5 slm recommended by the manufacturer. The ratio of N_2 to O_2 was chosen to be similar to the ratio in air for reliable comparison to the plasma operated in ambient air. Estimated values of RONS concentrations are summarized in Figure 7.

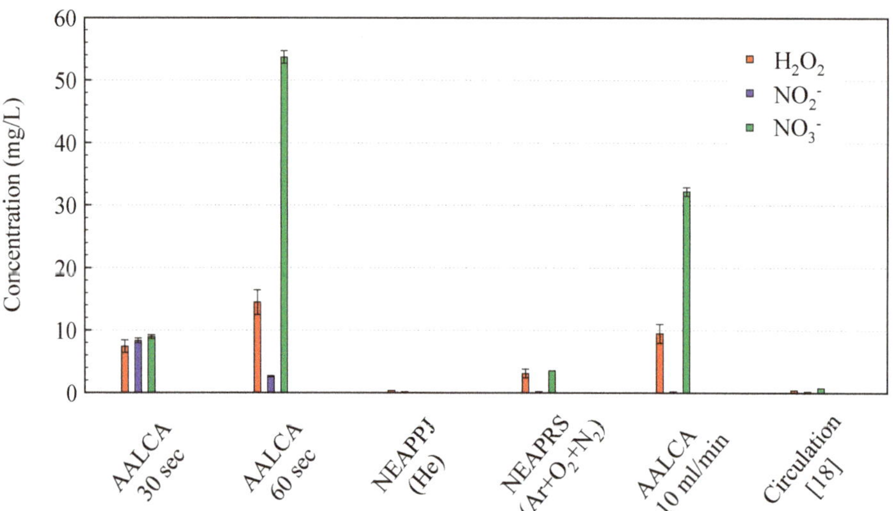

Figure 7. Comparison of concentrations of reactive oxygen and nitrogen species (RONS) in plasma-treated water (PTW) produced using treatment of 5 mL of DI water with AALCA, nonequilibrium atmospheric pressure radical source (NEAPRS), and nonequilibrium atmospheric pressure plasma jets (NEAPPJ); treatment of 10 mL/min flow of DI water with AALCA in the present work; and air plasma treatment of circulating liquid reported elsewhere [18].

It was found that the concentrations of H_2O_2, NO_2^-, and NO_3^- were 7.4, 8.3, and 8.9 mg/L, respectively, after 30 s of irradiation by AALCA; moreover, the concentrations of H_2O_2 and NO_3^- further increased to 14.5 and 53.7 mg/L, respectively, when the concentration of NO_2^- decreased to 2.7 mg/L with an increase of the treatment duration to 60 s. Owing to the longer treatment duration, NO_2^- might convert to more stable NO_3^-, resulting in a change of concentration ratios. The observed concentrations were two orders of magnitude higher than those of 5 mL DI water treated for 1 min with commercially available NEAPPJ operated with He (0.37 mg/L of H_2O_2, 0.18 mg/L of NO_2^-, and 0.06 mg/L of NO_3^-). Additionally, considerably small concentrations of RONS in water treated using NEAPPJ (up to 890 µg/L of H_2O_2, up to 180 µg/L of NO_2^-, and up to 410 µg/L of NO_3^-) [20–22,45,46] and custom plasma jets operated with air or addition of air (up to 9.3 mg/L of H_2O_2, up to 4 mg/L of NO_2^-, and up to 5.5 mg/L of NO_3^-) [14,28,47,48] were reported elsewhere, confirming the high efficiency of the developed AALCA as compared with commonly used nonequilibrium plasma jets.

Considering the low power consumption of conventional NEAPPJ compared with AALCA, a more reliable comparison is to NEAPRS, which has a power consumption of about 50 W, which is comparable to AALCA. After treatment of 5 mL of DI water with NEAPRS for 1 min, the concentrations of H_2O_2, NO_2^-, and NO_3^- were 3.1, 0.2, and 3.6 mg/L, respectively, which was one order of magnitude lower than the 5 mL of DI water treated with AALCA for 1 min.

3.4. Inactivation of E. coli

There are numerous reports about cancer therapy, disinfection, and inactivation of fungi using PTW, and the high concentration of RONS in PTW produced by AALCA looks promising for many such applications. To confirm the bactericidal effect of direct treatment using AALCA, a suspension containing *E. coli* in a concentration of 10^7 mL^{-1} was treated with plasma. The number of surviving *E. coli* was estimated using the colony counting method and the results are presented in Figure 8.

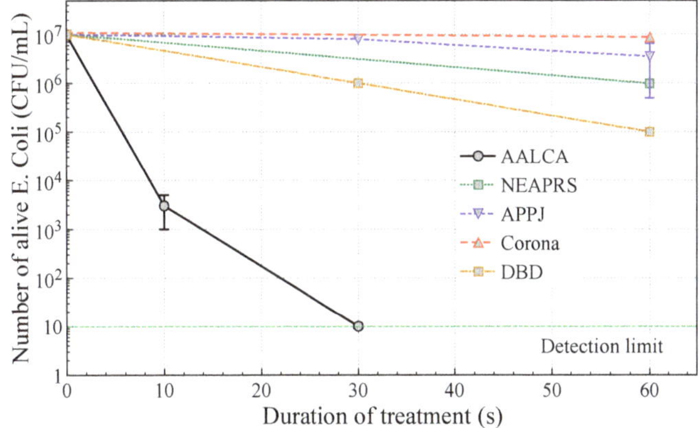

Figure 8. Number of surviving *E. coli* after treatment by AALCA, NEAPRS, NEAPPJ [43], dielectric barrier discharges (DBD) [49] or corona [50], discharges for various time of irradiation.

It was found that the number of surviving *E. coli* was about 2×10^3 mL^{-1} after direct plasma treatment for 10 s and decreased to the colony counting method detection limit of 10 mL^{-1} after 30 s, showing a 6-log reduction in the concentration of bacteria. As shown in Figure 8, the bactericidal effect of treatment with AALCA was significantly higher than the bactericidal effect of treatment with NEAPRS and APPJ, corona, or DBD discharges reported elsewhere [9,43,49–56]. It was confirmed by thermographic observation that the suspension was heated by irradiation to temperatures up to 38 °C, which is below the temperature harmful for *E. coli* reported elsewhere, suggesting that heating of the sample was not involved in the sterilization process [57]. On the other hand, the low pH value of about

3.4 after treatment of DI water by AALCA for 30 s could be effective for inactivation of bacteria [43]. To investigate the effect of temperature and pH on killing the bacteria, *E. coli* was suspended for 30 s in DI water heated to 40 °C (to ensure that the temperature of the suspension was higher than the temperature of the sample treated by AALCA for 30 s), citrate-phosphate buffer solution with a pH of 3.3 (similar to the pH observed for DI water treated by AALCA for 30 s), and citrate-phosphate buffer solution with a pH of 3.3 heated to 40 °C. Then, the number of surviving bacteria was estimated using the colony counting method. It was found that the concentration of bacteria did not change after suspension for 30 s in DI water heated to 40 °C or citrate-phosphate buffer solution with a pH of 3.3; however, a 1-log reduction in the concentration of *E. coli* was observed in the case of suspension in the citrate-phosphate buffer solution with a pH of 3.3 that was heated to 40 °C, indicating that the combination of heating the sample and reducing the pH value during AALCA treatment could contribute to observed bactericidal effect.

On the other hand, the high concentration of RONS in the DI water treated by AALCA (Figure 7) could be the origin of the observed bactericidal effect. To investigate the effect of H_2O_2, NO_2^-, and NO_3^- on sterilization of the bacteria, a solution containing 7 mg/L of H_2O_2, 8 mg/L of NO_2^-, and 9 mg/L of NO_3^- (similar to the concentrations observed after treatment by AALCA for 30 s) was prepared using chemical reagents, and *E. coli* was suspended for 30 s in the prepared solution. It was found by the colony counting method that the concentration of *E. coli* did not change after suspension for 30 s in the solution containing H_2O_2, NO_2^-, and NO_3^-, indicating that the high concentration of long-living RONS did not contribute to the strong bactericidal effect observed after 30 s of treatment. However, in the case of direct treatment with AALCA, not only long-living RONS could be involved in the bactericidal effect but also short-living RONS (e.g., $ONOO^-$ and $HOO\bullet$), the strong electric field at the stage of corona and repeating spark discharges, and ultraviolet emissions, as it was reported elsewhere [18,41,58,59].

On the other hand, a long-term bactericidal effect produced by long-living RONS in PTW was reported elsewhere, which suggests a strong long-term bactericidal effect of PTW produced by AALCA owing to its high concentration of RONS [31,32,43]. To investigate the bactericidal effect of PTW, 5 mL of DI water was treated with AALCA for 1 min at an input power of 28 W. Then, 0.3 mL of the *E. coli* suspension was mixed with 2.7 mL of the produced PTW and incubated for various times. After incubation, the number of surviving bacteria was estimated using the colony counting method. The estimated values of *E. coli* after various durations of incubation are presented in Figure 9. It could be clearly observed that the concentration of *E. coli* gradually decreased as the incubation time increased after suspension of *E. coli* in DI water treated with AALCA and reached the detection limit of 10 CFU/mL at 60 min, showing a 6-log reduction in the concentration of bacteria. Considering that the concentration of *E. coli* did not change in the control sample of *E. coli* suspended in pristine DI water even after 24 h of incubation, it could be concluded that the DI water treated with AALCA had a strong bactericidal effect that originated from the high concentration of RONS and the low pH level [31,32,43].

Treatment of liquid using AALCA could be performed irrespective of its electrical conductivity; therefore, further reduction of production costs is possible by using tap or spring water. Tap water was treated with AALCA under the same conditions and the bactericidal effect of plasma-treated tap water was investigated in the same way as for DI water. It can be clearly observed in Figure 9 that tap water treated by AALCA also had a bactericidal effect, and the concentration of bacteria reached the detection limit of 10 CFU/mL after 90 min, showing a 6-log reduction in the concentration of bacteria. Considering that the concentration of bacteria in the control sample of E. coli suspended in pristine tap water did not change after 90 min of incubation and reduced by 1 log only after 24 h of incubation, it can be concluded that, like with DI water, tap water treated with AALCA has a strong bactericidal effect that originates mostly from the high concentration of RONS. The difference in the bactericidal effect can be explained by the complex composition of tap water. Some components of tap water may interfere with chemical reactions in PTW and reduce the amount of species responsible for the bactericidal effect. Additionally, the pH value of the tap water was higher than that of DI water

(Figure 4), which may have reduced the bactericidal efficacy of tap water treated with AALCA as compared with DI water [33,43].

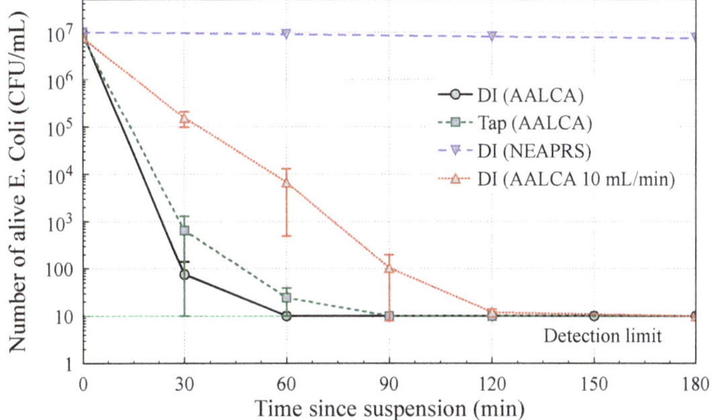

Figure 9. Dependency of the number of surviving bacteria on the duration of incubation after mixture of *E. coli* suspension with DI or tap water treated with AALCA, DI water treated with NEAPRS, or DI water treated with AALCA at a 10 mL/min flow.

The observed bactericidal effect of both DI and tap water, together with its low production cost, makes PTW produced by AALCA appear to be promising as a low-cost disinfectant which could be used in biomedical applications. However, the composition and properties of tap water can vary strongly depending on the area of use. Therefore, more comprehensive investigations are needed for the practical use of plasma-treated tap water in various locations.

3.5. Treatment of Flowing and Circulating Liquid

The evaporation of liquid reported above might be a problem for AALCA treatment; however, considering the high efficiency of RONS production and the moderate temperature of the liquid treated using the proposed method, AALCA treatment looks promising for practical PTW use. A possible approach to reduce the production cost of PTW and ameliorate the problems associated with heating and evaporation of the liquid is treatment by continuous flow of liquid. Treatment of a flow of DI water was performed using the same setup and two peristatic pumps, as shown in Figure 10. Pristine DI water was introduced to a dish at a flow rate of 10.25 mL/min using the first pump and PTW was evacuated from the dish at a flow rate of 10 mL/min using the second pump. The flow allowed for reducing the temperature of the liquid during treatment. It was found by thermographic measurements (Figure 10) that the temperature of the treated water was about 45 °C in the area of contact with plasma after 5 min of operation, which was significantly lower than the case of the 5 mL sample treatment. Figure 10 shows a thermographic image of plastic tubes with water on the input and output of the plasma treatment device. The temperature of the water increased from about 25 °C to about 30 °C during the treatment (similar results were confirmed by measurements using a thermocouple). The use of flow allows for a temperature of the produced PTW below 37 °C, which is safe for the human body and could be used in medical applications [47,60]. Additionally, the difference in the flow rate allowed for compensation of the evaporation of the liquid and retention of the same level of liquid inside the dish, ensuring the same discharge gap and parameters of the plasma during treatment.

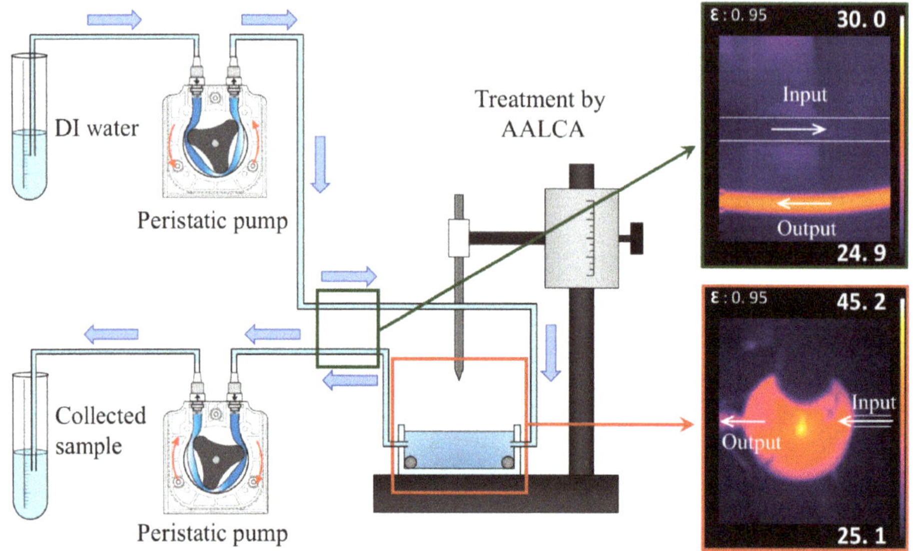

Figure 10. Schematic of experimental setup for treatment of continuous flow of liquid, thermographic image of flowing DI water treated with AALCA (right, bottom), and comparison of water temperatures in tubes before and after treatment (right, top).

The treated water was collected from the output of the second pump and concentrations of RONS in the collected PTW were estimated using HPLC and ion chromatography. Estimated concentrations of H_2O_2, NO_2^-, and NO_3^- in the produced PTW were 9.6, 0.2, and 32.2 mg/L, respectively, as shown in Figure 7. The pH value of the produced PTW was about 3.25. The production rate of RONS was comparable to the treatment of stationary samples of 5 mL. Despite the reduced efficiency of production of NO_2^-, the total amount of RONS in the case of the flow of water was still two orders of magnitude higher than the amounts observed for the case of conventional NEAPPJ or DBD [14,20,32,43]. Zhou et al. reported PTW production by cold plasma treatment at ambient air of 50 mL of circulating liquid at a flow rate of 200 mL/min [18]. It is difficult to perform a direct comparison of results considering the difference in the flow ratio of the liquid and the experimental setup. It would take 5 min to produce 50 mL of PTW using AALCA with a 10 mL/min flow of the water used in the present work; therefore, the results of the present work were compared to the results reported by Zhou et al. for 5 min of treatment (0.4 of H_2O_2, 0.2 of NO_2^-, and 0.75 of NO_3^-) [18]. It can be clearly observed in Figure 7 that treatment with AALCA is more effective than the method reported by Zhou et al., and treatment of a liquid flow can be applied to produce a large volume of PTW. However, the concentration of NO_2^- in the produced PTW was lower than that of the fast treatment of stationary liquids with small volumes, and the ratio of concentrations of NO_2^- and NO_3^- depends on the flow rate, duration of treatment, and discharge parameters.

To investigate the bactericidal effect of treatment of flowing liquid using AALCA, an *E. coli* suspension with a concentration of bacteria of 10^7 mL^{-1} was treated with AALCA at an input power of 28 W and a flow rate of 10 mL/min. It was found by the colony counting method that the concentration of bacteria decreased to the detection limit of 10 CFU/mL after treatment by AALCA at a flow rate of 10 mL/min, showing a 6-log reduction in the concentration of bacteria. This result demonstrates that treatment of a continuous flow of liquid by AALCA is applicable to disinfection of larger volumes of liquids.

To investigate the bactericidal effect of PTW produced by treatment of flowing liquid using AALCA, DI water was treated with AALCA at an input power of 28 W and a flow rate of 10 mL/min and collected in a clean vial. Then, 0.3 mL of the *E. coli* suspension was mixed with 2.7 mL of collected PTW and incubated for various times. After incubation, the number of surviving bacteria was estimated using the colony counting method. The estimated values of *E. coli* after various durations of incubation are presented in Figure 9. It can be clearly observed that the concentration of bacteria gradually decreased as the incubation time increased after suspension of *E. coli* to produce PTW, and it reached the detection limit of 10 CFU/mL at 180 min, showing a 6-log reduction in the concentration of bacteria. Considering that the concentration of *E. coli* did not change in the control sample of *E. coli* suspended in pristine DI water, it can be concluded that PTW produced by treatment of a continuous flow of liquid with AALCA has a strong bactericidal effect originating from the high concentration of RONS. The difference in bactericidal effect observed for the 5 mL sample and the flowing DI water can be explained by the difference in concentrations of RONS and the higher pH. In the case of flowing DI water, the concentration of NO_2^- was about 10 times lower and the pH was higher than the 5 mL of DI water treated by AALCA. Moreover, in the case of a flowing liquid, the concentrations of short-living reactive species could be different, resulting in a reduced bactericidal effect.

We found in this work that AALCA is effective at generating RONS and features low production and operation costs, which helps to overcome a number of problems in existing systems for plasma treatment of liquids. However, a major drawback of AALCA is the heating and evaporation of the liquid. We demonstrated that these problems can be solved by treatment of a continuous flow of liquid, which avoids changes in the discharge gap by evaporation and achieves a temperature safe for the human body that is below 37 °C after treatment [60]. The low temperature and high concentration of RONS in PTW produced by treatment of flowing liquid looks promising for biomedical and agricultural applications. Furthermore, production of PTW using a continuous flow of liquid could be easily scaled to large volumes, which looks promising for water disinfection and purification, as well as other applications requiring low-cost treatment of large volumes of liquid. Treatment of a liquid flow using AALCA allows for the low-cost production of PTW and solves several problems associated with using a stationary solution, however, further tuning of the discharge conditions and development of a custom pumping system is necessary for precise control of RONS concentrations in the produced PTW.

Additionally, the stable generation of AALCA was confirmed using liquids with various conductivities, such as DI water, tap water, ethanol, propylene glycol, phosphate buffer, phosphate saline buffer, and phosphate buffer containing amino acids. The possibility of generating such a type of plasma on the surface of a liquid irrespective of the electrical conductivity could be essential for various biomedical and agricultural applications, where buffer solutions with various conductivities are commonly used to control the pH level.

Moreover, the possibility of treating liquid precursors of various conductivities by AALCA looks promising for the synthesis of nanoparticles and nanocarbons [1,8,10,11].

4. Conclusions

In this work, we developed a compact tool for irradiation of liquids by a low-current arc at ambient air. It was possible to generate a stable discharge, even when using liquids that had low electrical conductivity, as a sequence of corona, repeating spark, and low-current arc discharges. Moreover, the possibility of treating a continuous flow of liquid was confirmed. It was found that the concentration of RONS after treatment using an ambient air low-current arc was two orders of magnitude higher than that of PTW produced using conventional APPJs or DBD plasmas under similar treatment conditions. The strong bactericidal effect of treatment with AALCA and the produced PTW was confirmed by survival tests of *E. coli*. Additionally, we demonstrated a method to overcome the overheating problem and increase the volume of produced PTW by treatment of a continuous flow of liquid. This work is a step towards the practical use of plasma-treated liquids in biomedical and agricultural applications.

Author Contributions: Experiments, investigation, and writing—review and editing, V.G.; survival test of *E. coli*, N.I.; resources, M.H. (Masaru Hori); funding acquisition, M.H. (Mineo Hiramatsu); validation and supervision, M.I.

Funding: This research was funded by the MEXT-Supported Program for the Strategic Research Foundation at Private Universities, grant number S1511021.

Conflicts of Interest: The authors declare no conflict of interest.

References

1. Hagino, T.; Kondo, H.; Ishikawa, K.; Kano, H.; Sekine, M.; Hori, M. Ultrahigh-speed synthesis of nanographene using alcohol in-liquid plasma. *Appl. Phys. Express* **2012**, *5*, 035101. [CrossRef]

2. Chen, Z.; Lin, L.; Cheng, X.; Gjika, E.; Keidar, M. Effects of cold atmospheric plasma generated in deionized water in cell cancer therapy. *Plasma Process. Polym.* **2016**, *13*, 1151–1156. [CrossRef]

3. Gamaleev, V.; Iwata, N.; Oh, J.S.; Hiramatsu, M.; Ito, M. Development of an Ambient Air Flow Rotating Arc Jet for Low-Temperature Treatment. *IEEE Access* **2019**, *7*, 93100–93107. [CrossRef]

4. Iwata, N.; Gamaleev, V.; Hashizume, H.; Oh, J.; Ohta, T.; Ishikawa, K.; Hori, M.; Ito, M. Simultaneous achievement of antimicrobial property and plant growth promotion using plasma-activated benzoic compound solution. *Plasma Process. Polym.* **2019**, e1900023. [CrossRef]

5. Zhang, J.; Kwon, T.; Kim, S.; Jeong, D. Plasma farming: Non-Thermal dielectric barrier discharge plasma technology for improving the growth of soybean sprouts and chickens. *Plasma* **2018**, *1*, 285–296. [CrossRef]

6. Kučerová, K.; Henselová, M.; Slováková, L.; Hensel, K. Effects of plasma activated water on wheat: Germination, growth parameters, photosynthetic pigments, soluble protein content, and antioxidant enzymes activity. *Plasma Process. Polym.* **2019**, *16*, 1800131. [CrossRef]

7. Li, Y.; Wang, J.; Zhang, J.; Fang, J. In vitro studies of the antimicrobial effect of non-thermal plasma-activated water as a novel mouthwash. *Eur. J. Oral Sci.* **2017**, *125*, 463–470. [CrossRef]

8. Amano, T.; Kondo, H.; Ishikawa, K.; Tsutsumi, T.; Takeda, K.; Hiramatsu, M.; Sekine, M.; Hori, M. Rapid growth of micron-sized graphene flakes using in-liquid plasma employing iron phthalocyanine-added ethanol. *Appl. Phys. Express* **2018**, *11*, 015102. [CrossRef]

9. Shaw, P.; Kumar, N.; Kwak, H.S.; Park, J.H.; Uhm, H.S.; Bogaerts, A.; Choi, E.H.; Attri, P. Bacterial inactivation by plasma treated water enhanced by reactive nitrogen species. *Sci. Rep.* **2018**, *8*, 11268. [CrossRef]

10. Saito, N.; Bratescu, M.A.; Hashimi, K. Solution plasma: A new reaction field for nanomaterials synthesis. *Jpn. J. Appl. Phys.* **2018**, *57*, 0102A4. [CrossRef]

11. Gamaleev, V.; Kajikawa, K.; Takeda, K.; Hiramatsu, M. Investigation of nanographene produced by in-liquid plasma for development of highly durable polymer electrolyte fuel cells. *C* **2018**, *4*, 65. [CrossRef]

12. Ito, M.; Oh, J.S.; Ohta, T.; Shiratani, M.; Hori, M. Current status and future prospects of agricultural applications using atmospheric-pressure plasma technologies. *Plasma Process. Polym.* **2018**, *15*, e1700073. [CrossRef]

13. Judée, F.; Simon, S.; Bailly, C.; Dufour, T. Plasma-activation of tap water using DBD for agronomy applications: Identification and quantification of long lifetime chemical species and production/consumption mechanisms. *Water Res.* **2018**, *133*, 47–59. [CrossRef]

14. Zhang, X.; Zhou, R.; Bazaka, K.; Liu, Y.; Zhou, R.; Chen, G.; Chen, Z.; Qinghuo, L.; Yang, S.; Ostrikov, K.K. Quantification of plasma produced OH radical density for water sterilization. *Plasma Process. Polym.* **2018**, *15*, e1700241. [CrossRef]

15. Zhang, Z.; Shen, J.; Cheng, C.; Xu, Z.; Xia, W. Generation of reactive species in atmospheric pressure dielectric barrier discharge with liquid water. *Plasma Sci. Technol.* **2018**, *20*, 44009. [CrossRef]

16. Tang, Q.; Jiang, W.; Cheng, Y.; Lin, S.; Lim, T.M.; Xiong, J. Generation of reactive species by gas-phase dielectric barrier discharges. *Ind. Eng. Chem. Res.* **2011**, *50*, 9839–9846. [CrossRef]

17. Xu, H.; Liu, D.; Wang, W.; Liu, Z.; Guo, L.; Rong, M.; Kong, M.G. Investigation on the RONS and bactericidal effects induced by He + O2 cold plasma jets: In open air and in an airtight chamber. *Phys. Plasmas* **2018**, *25*, 113506. [CrossRef]

18. Zhou, R.; Zhou, R.; Prasad, K.; Fang, Z.; Speight, R.; Bazaka, K.; Ostrikov, K. Cold atmospheric plasma activated water as a prospective disinfectant: The crucial role of peroxynitrite. *Green Chem.* **2018**, *20*, 5276–5284. [CrossRef]

19. Oh, J.S.; Kakuta, M.; Furuta, H.; Akatsuka, H.; Hatta, A. Effect of plasma jet diameter on the efficiency of reactive oxygen and nitrogen species generation in water. *Jpn. J. Appl. Phys.* **2016**, *55*, 06HD01. [CrossRef]

20. Chauvin, J.; Judée, F.; Yousfi, M.; Vicendo, P.; Merbahi, N. Analysis of reactive oxygen and nitrogen species generated in three liquid media by low temperature helium plasma jet. *Sci. Rep.* **2017**, *7*, 4562. [CrossRef]

21. Ogawa, K.; Oh, J.; Gaur, N.; Hong, S.; Kurita, H.; Mizuno, A.; Hatta, A.; Short, R.D.; Ito, M.; Szili, E.J. Modulating the concentrations of reactive oxygen and nitrogen species and oxygen in water with helium and argon gas and plasma jets. *Jpn. J. Appl. Phys.* **2019**, *58*, SAAB01. [CrossRef]

22. Kim, D.Y.; Kim, S.J.; Joh, H.M.; Chung, T.H. Characterization of an atmospheric pressure plasma jet array and its application to cancer cell treatment using plasma activated medium. *Phys. Plasmas* **2018**, *25*, 073505. [CrossRef]

23. Zhou, R.; Zhou, R.; Zhuang, J.; Zong, Z.; Zhang, X.; Liu, D.; Bazaka, K.; Ostrikov, K. Interaction of atmospheric-pressure air microplasmas with amino acids as fundamental processes in aqueous solution. *PLoS ONE* **2016**, *11*, e0155584. [CrossRef]

24. Schmidt, M.; Hahn, V.; Altrock, B.; Gerling, T.; Gerber, I.C.; Weltmann, K.; Woedtke, T. Von Plasma-Activation of Larger Liquid Volumes by an Inductively-Limited Discharge for Antimicrobial Purposes. *Appl. Sci.* **2019**, *9*, 2150. [CrossRef]

25. Gharagozalian, M.; Dorranian, D.; Ghoranneviss, M. Water treatment by the AC gliding arc air plasma. *J. Theor. Appl. Phys.* **2017**, *11*, 171–180. [CrossRef]

26. Starek, A.; Pawłat, J.; Chudzik, B.; Kwiatkowski, M.; Terebun, P.; Sagan, A.; Andrejko, D. Evaluation of selected microbial and physicochemical parameters of fresh tomato juice after cold atmospheric pressure plasma treatment during refrigerated storage. *Sci. Rep.* **2019**, *9*, 8407. [CrossRef]

27. Hasan, M.I.; Walsh, J.L. The generation and transport of reactive nitrogen species from a low temperature atmospheric pressure air plasma source. *Phys. Chem. Chem. Phys.* **2018**, *20*, 28499–28510. [CrossRef]

28. Joh, H.M.; Choi, J.Y.; Kim, S.J.; Chung, T.H. Effects of the pulse width on the reactive species production and DNA damage in cancer cells exposed to atmospheric pressure microsecond-pulsed helium plasma jets. *AIP Adv.* **2019**, *26*, 053509. [CrossRef]

29. Kaushik, N.K.; Ghimire, B.; Li, Y.; Adhikari, M.; Veerana, M.; Kaushik, N.; Jha, N.; Adhikari, B.; Lee, S.J.; Masur, K.; et al. Biological and medical application of plasma-activated media, water and solutions. *Biol. Chem.* **2018**, *400*, 39–62. [CrossRef]

30. Lo, C.; Ziuzina, D.; Los, A.; Boehm, D.; Palumbo, F.; Favia, P.; Tiwari, B.; Bourke, P.; Cullen, P.J. Plasma activated water and airborne ultrasound treatments for enhanced germination and growth of soybean. *Innov. Food Sci. Emerg. Technol.* **2018**, *49*, 13–19. [CrossRef]

31. Iwata, N.; Gamaleev, V.; Oh, J.S.; Ohta, T.; Hori, M.; Ito, M. Investigation on the long-term bactericidal effect and chemical composition of radical-activated water. *Plasma Process. Polym.* **2019**, e1900055. [CrossRef]

32. Traylor, M.J.; Pavlovich, M.J.; Karim, S.; Hait, P.; Sakiyama, Y.; Clark, D.S.; Graves, D.B. Long-term antibacterial efficacy of air plasma-activated water. *J. Phys. D Appl. Phys.* **2011**, *44*, 472001. [CrossRef]

33. Kim, H.; Wright, K.C.; Hwang, I.; Lee, D.; Rabinovich, A.; Fridman, A.; Cho, Y.I. Effects of H2O2 and low pH produced by gliding Arc discharge on the inactivation of escherichia Coli in Water. *Plasma Med.* **2011**, *1*, 295–307. [CrossRef]

34. Li, L.; Zhang, H.; Yan, J.; Li, X.; Wang, W.; Tu, X. Warm plasma activation of CO2 in a rotating gliding arc discharge reactor. *J. CO2 Util.* **2018**, *27*, 472–479. [CrossRef]

35. Gamaleev, V.; Furuta, H.; Hatta, A. Detection of metal contaminants in seawater by spectral analysis of microarc discharge. *Jpn. J. Appl. Phys.* **2018**, *57*, 0102B8. [CrossRef]

36. Gamaleev, V.; Furuta, H.; Hatta, A. Atomic emission spectroscopy of microarc discharge in sea water for on-site detection of metals. *IEEE Trans. Plasma Sci.* **2019**, *47*, 1841–1850. [CrossRef]

37. Gamaleev, V.; Furuta, H.; Hatta, A. Generation of micro-arc discharge plasma in highly pressurized seawater. *Appl. Phys. Lett.* **2018**, *113*, 214102. [CrossRef]

38. Chanan, N.; Kusumandari; Saraswati, T.E. Water Treatment Using Plasma Discharge with Variation of Electrode Materials. *IOP Conf. Ser. Mater. Sci. Eng.* **2018**, *333*, 012025. [CrossRef]

39. Bruggeman, P.J.; Kushner, M.J.; Locke, B.R.; Gardeniers, J.G.E.; Graham, W.G.; Graves, D.B.; Hofman-Caris, R.C.H.M.; Maric, D.; Reid, J.P.; Ceriani, E.; et al. Plasma-liquid interactions: A review and roadmap. *Plasma Sources Sci. Technol.* **2016**, *25*, 053002. [CrossRef]

40. Miyamoto, K.; Ikehara, S.; Takei, H.; Akimoto, Y.; Sakakita, H.; Ishikawa, K.; Ueda, M.; Ikeda, J.; Yamagishi, M.; Kim, J.; et al. Red blood cell coagulation induced by low-temperature plasma treatment. *Arch. Biochem. Biophys.* **2016**, *605*, 95–101. [CrossRef]

41. Iseki, S.; Hashizume, H.; Jia, F.; Takeda, K.; Ishikawa, K.; Ohta, T.; Ito, M.; Hori, M. Inactivation of Penicillium digitatum Spores by a High-Density Ground-State Atomic Oxygen-Radical Source Employing an Atmospheric-Pressure Plasma. *Appl. Phys. Express* **2011**, *4*, 116201. [CrossRef]

42. Bruggeman, P.; Ribel, E.; Maslani, A.; Degroote, J.; Malesevic, A.; Rego, R.; Vierendeels, J.; Leys, C. Characteristics of atmospheric pressure air discharges with a liquid cathode and a metal anode. *Plasma Sources Sci. Technol.* **2008**, *17*, 025012. [CrossRef]

43. Ikawa, S.; Kitano, K.; Hamaguchi, S. Effects of pH on bacterial inactivation in aqueous solutions due to low-temperature atmospheric pressure plasma application. *Plasma Process. Polym.* **2010**, *7*, 33–42. [CrossRef]

44. Bibinov, N.K.; Fateev, A.A.; Wiesemann, K. On the influence of metastable reactions on rotational temperatures in dielectric barrier discharges in He-N2mixtures. *J. Phys. D Appl. Phys.* **2001**, *34*, 1819–1826. [CrossRef]

45. Oh, J.; Szili, E.J.; Ogawa, K.; Short, R.D.; Ito, M.; Furuta, H.; Hatta, A. UV-vis spectroscopy study of plasma-activated water: Dependence of the chemical composition on plasma exposure time and treatment distance. *Jpn. J. Appl. Phys.* **2018**, *57*, 0102B9. [CrossRef]

46. Oh, J.-S.; Szili, E.J.; Gaur, N.; Hong, S.-H.; Furuta, H.; Kurita, H.; Mizuno, A.; Hatta, A.; Short, R.D. How to assess the plasma delivery of RONS into tissue fluid and tissue. *J. Phys. D Appl. Phys.* **2016**, *49*, 304005. [CrossRef]

47. Xaubet, M.; Baudler, J.S.; Gerling, T.; Giuliani, L.; Minotti, F.; Grondona, D.; Von Woedtke, T.; Weltmann, K.D. Design optimization of an air atmospheric pressure plasma-jet device intended for medical use. *Plasma Process. Polym.* **2018**, *15*, e1700211. [CrossRef]

48. Pawłat, J.; Terebun, P.; Kwiatkowski, M.; Tarabová, B.; Kovaľová, Z.; Kučerová, K.; Machala, Z.; Janda, M.; Hensel, K. Evaluation of Oxidative Species in Gaseous and Liquid Phase Generated by Mini-Gliding Arc Discharge. *Plasma Chem. Plasma Process.* **2019**, *39*, 627–642. [CrossRef]

49. Hong, E.J.; Shin, J.H.; Ji, S.H.; Ahn, J.H.; Kim, Y.J.; Choi, E.H.; Ki, S.H. Inactivation of Escherichia coli and Staphylococcus aureus on contaminated perilla leaves by Dielectric Barrier Discharge (DBD) plasma treatment. *Arch. Biochem. Biophys.* **2018**, *643*, 32–41. [CrossRef]

50. Xu, Z.; Cheng, C.; Shen, J.; Lan, Y.; Hu, S.; Han, W.; Chu, P.K. In vitro antimicrobial effects and mechanisms of direct current air-liquid discharge plasma on planktonic Staphylococcus aureus and Escherichia coli in liquids. *Bioelectrochemistry* **2018**, *121*, 125–134. [CrossRef]

51. Machala, L.; Tarabova, B.; Sersenova, J.; Hensel, K. Chemical and antibacterial effects of plasma activated water: Correlation with gaseous and aqueous reactive oxygen and nitrogen species. *J. Phys. D Appl. Phys.* **2019**, *52*, 034002. [CrossRef]

52. Dors, M.; Metel, E.; Mizeraczyk, J.; Marotta, E. Pulsed corona discharge in water for coli bacteria inactivation. *Int. J. Plasma Environ. Sci. Technol.* **2008**, *2*, 34–37. [CrossRef]

53. Du, C.; Tang, J.; Mo, J.; Ma, D.; Wang, J.; Wang, K.; Zeng, Y. Decontamination of bacteria by gas-liquid gliding arc discharge: Application to escherichia coli. *IEEE Trans. Plasma Sci.* **2014**, *42*, 2221–2228. [CrossRef]

54. Han, L.; Patil, S.; Boehm, D.; Milosavljević, V.; Cullen, P.J.; Bourke, P. Mechanisms of inactivation by high-voltage atmospheric cold plasma differ for Escherichia coli and Staphylococcus aureus. *Appl. Environ. Microbiol.* **2016**, *82*, 450–458. [CrossRef]

55. Satoh, K.; MacGregor, S.J.; Anderson, J.G.; Woolsey, G.A.; Fouracre, R.A. Pulsed-plasma disinfection of water containing Escherichia coli. *Jpn. J. Appl. Phys.* **2007**, *46*, 1137–1141. [CrossRef]

56. Asadollahfardi, G.; Khandan, M.; Aria, S.H. Deactivation of Escherichia coli with different volumes in drinking water using cold atmospheric plasma. *Contrib. Plasma Phys.* **2019**, e201800106. [CrossRef]

57. Hutchison, M.L.; Thomas, D.J.I.; Avery, S.M. Thermal death of Escherichia coli O157: H7 in cattle feeds. *Lett. Appl. Microbiol.* **2007**, *44*, 357–363. [CrossRef]

58. Vermeulen, N.; Keeler, W.J.; Nandakumar, K.; Leung, K.T. The bactericidal effect of ultraviolet and visible light on Escherichia coli. *Biotechnol. Bioeng.* **2008**, *99*, 550–556. [CrossRef]
59. Machado, L.F.; Pereira, R.N.; Martins, R.C.; Teixeira, J.A.; Vicente, A.A. Moderate electric fields can inactivate Escherichia coli at room temperature. *J. Food Eng.* **2010**, *96*, 520–527. [CrossRef]
60. *DIN EN 60601-1-10:2016–04 Medical Electrical Equipment—Part 1: General Requirements for Basic Safety and Essential Performance*; Beuth Verlag: Berlin, Germany, 2016.

 *applied sciences*

Article

Plasma-Activation of Larger Liquid Volumes by an Inductively-Limited Discharge for Antimicrobial Purposes

Michael Schmidt [1,*], Veronika Hahn [1], Beke Altrock [1], Torsten Gerling [1], Ioana Cristina Gerber [2], Klaus-Dieter Weltmann [1] and Thomas von Woedtke [1,3]

[1] Leibniz Institute for Plasma Science and Technology, Felix-Hausdorff-Str. 2, 17489 Greifswald, Germany; veronika.hahn@inp-greifswald.de (V.H.); b.c.altrock@googlemail.com (B.A.); gerling@inp-greifswald.de (T.G.); weltmann@inp-greifswald.de (K.-D.W.); woedtke@inp-greifswald.de (T.v.W.)

[2] Faculty of Physics, Alexandru Ioan Cuza University of Iasi, 700506 Iaşi, Romania; cristinaioana.gerber@gmail.com

[3] Department of Hygiene and Environmental Medicine, Greifswald University Medicine, 17475 Greifswald, Germany

[*] Correspondence: M_Schmidt@inp-greifswald.de; Tel.: +49-3834-554-3866

Received: 26 April 2019; Accepted: 22 May 2019; Published: 27 May 2019

Abstract: A new configuration of a discharge chamber and power source for the treatment of up to 1 L of liquid is presented. A leakage transformer, energizing two metal electrodes positioned above the liquid, limits the discharge current inductively by utilizing the weak magnetic coupling between the primary and secondary coils. No additional means to avoid arcing (electric short-circuiting), e.g., dielectric barriers or resistors, are needed. By using this technique, exceeding the breakdown voltage leads to the formation of transient spark discharges, producing non-thermal plasma (NTP). These discharges effected significant changes in the properties of the treated liquids (distilled water, physiological saline solution, and tap water). Considerable concentrations of nitrite and nitrate were detected after the plasma treatment. Furthermore, all tested liquids gained strong antibacterial efficacy which was shown by inactivating suspended Escherichia coli and Staphylococcus aureus. Plasma-treated tap water had the strongest effect, which is shown for the first time. Additionally, the pH-value of tap water did not decrease during the plasma treatment, and its conductivity increased less than for the other tested liquids.

Keywords: inductively-limited discharge; plasma-treated water; tap water; antimicrobial activity

---

1. Introduction

In recent years, investigations of plasma-liquid interactions have become an important topic in plasma science and technology, not least being stimulated by the accelerating field of plasma medicine. Consequently, electrical discharges in combination with liquids have been extensively studied. Resulting from plasma-liquid interactions, very complex physical as well as chemical processes occur starting from gas phase chemistry via multiphase species transport, mass and heat transfer up to interfacial reactions and very complex liquid phase chemistry. In the case of low-temperature plasma sources working at atmospheric air conditions (cold atmospheric plasmas—CAP), this chemistry is mainly determined by reactive oxygen and nitrogen species transferred from the gas/plasma phase into the liquid phase or generated in the liquid phase by secondary reactions [1,2]. Several electrode configurations and modes of electrical operation for liquid treatment were investigated. Generally, these discharges can be ignited directly in the liquid [3], between one electrode above the liquid and the surface of the liquid itself serving as the counter electrode [4–6], and in close vicinity to the water surface, e.g., by a dielectric barrier discharge [7] or with a gliding arc discharge [8]. Configurations using plasma jets [9] or high-voltage-fed hollow needles [10] and electro-spraying were also investigated [11]. The discharges can be energized with DC, AC, or pulsed high voltage [4,12,13].

Besides the various ways of operating these discharges, their physical properties like breakdown voltage, discharge current and morphology, gas temperature, and light emission are discussed. As liquids, different kinds of water were used (distilled water, saline solution, buffered solutions, or tap water). For comprehensive summaries of the broad variety of discharges in combination with liquids, please refer to [1,14,15]. Regarding the treatment of water, changes of its properties have been examined by many researchers [6–9,16–19]. In all cases of discharges operated in contact with or in close vicinity to water, acidification (except for tap water) and nitration was observed. Additionally, it was described that chemical contaminations, e.g., with pharmaceuticals, in the water were decomposed over the course of the plasma treatment [20–22]. Further experiments revealed that the plasma-treated water became antimicrobial [5,7,18]. In these experiments, bacteria such as *Staphylococcus aureus* and *Escherichia coli* were significantly inactivated. Several tests were performed with bacterial suspensions, i.e., the microorganisms to be inactivated were present during plasma treatment of the liquid (direct plasma treatment). However, it was also demonstrated that plasma-treated liquids show antibacterial activity if the microorganisms were added immediately after plasma treatment of the liquid (indirect plasma treatment) [18,23]. Traylor et al. [24] have shown that distilled water which was plasma treated for 20 min retained its antibacterial activity for 7 days. Hänsch et al. [18] discussed that the plasma-generated species hydrogen peroxide and nitrite decrease but are still present 1.5 h after the treatment was finished. According to the present state of knowledge, reactive oxygen and nitrogen species (ROS and RNS) play an important role in plasma-mediated inactivation of microorganisms [25]. These findings have given rise to the conclusion that plasma treatment of aqueous liquids might be useful to produce biologically-active formulations for several applications in the medical field like disinfectants or antiseptics [26,27].

Many of the aforementioned experiments were performed with rather small volumes of water, i.e., in the range of milliliters. For the characterization of the plasma itself and the antimicrobial properties, as well as the chemistry of the liquids, this volume is sufficient. To enable relevant conditions for applications using plasma-treated liquid, larger volumes are required. Therefore, we designed a discharge configuration for the activation of liquid volumes up to 1 L. In order to ensure easy operation and handling, we used a pin-to-liquid design. The limiting of the discharge current turned out to be the most challenging part of the development, because dielectric barriers on the electrodes require very high operating voltages, whereas current limiting resistors waste too much electrical energy by heating. As a solution, we found inductive limiting of the discharge current by a choke. This resulted in the use of a leakage transformer, which will be discussed later in detail.

2. Materials and Methods

2.1. Electrical and Optical Investigations

For the investigations, two different experimental setups were designed. The experimental setup used for the assessment of the discharge parameters consisted of two electrodes made of stainless steel (diameter 4 mm, length 32 mm). These electrodes were fixed to an electrode holder and positioned about 3 mm above the surface of the liquid in a beaker with a volume of 250 mL (Figure 1). The electrode holder was a round shape disk made of polycarbonate with a diameter of 120 mm and a thickness of 80 mm. The electrodes were connected to the output transformer of a commercial high-voltage source (Neon-Pro NP 10000-30) operating the discharge at a frequency of approximately 25 kHz and ignition voltages of about 2–3 kV (amplitude). The high voltage source was driven by a variac (RFT LTS 002) connected directly to the net. Details of the electrical parameters are given in the electrical characterization section.

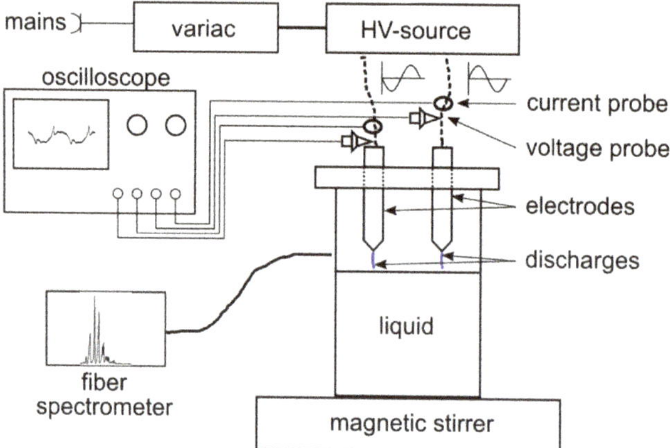

Figure 1. Experimental setup.

The beaker with the electrode holder was placed on a magnetic stirrer (Thermo Scientific Cimarec i Mono) and the water was stirred permanently. The electrical data were monitored with high voltage probes (Tektronix P6015A) and current probes (Pearson 2877) and recorded with a digital oscilloscope (Tektronix DPO 4104, Tektronix, Beaverton, OR, US). The power dissipated into the high voltage source was measured by a voltage meter (PCE-PA 6000). The discharges were optically investigated with a fiber spectrometer (Ocean Optics HR4000, Ocean Optics, Largo, FL, US) via an orifice in the beaker.

For the plasma-treatment of water, the electrode configuration described above was doubled, as described in [28]. Thus, four electrodes were positioned with a holder on top of a beaker with a volume of 0.5 L or 1 L. Pictures of the discharge configurations (Figure 2) were taken with a digital camera (Canon EOS 70D with a macro lens Tamron XR Di II). Tap water, physiological saline solution (distilled water with 0.85% NaCl), and distilled water were treated for up to 30 min.

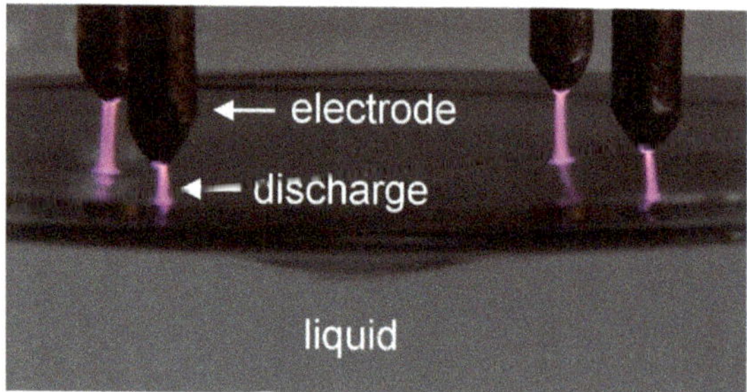

Figure 2. Photograph of the 4-electrode configuration for treatment of up to 1 L water [28].

2.2. Investigation of Properties of the Treated Water

The pH-value, the conductivity, and the temperature were measured with a multiparameter meter (Hanna Instruments HI 9828, Vöhringen, Germany. Nitrogen species, namely nitrite and nitrate, were detected by ion chromatography using a Dionex ICS 5000 system (Thermo Scientific, Dreieich, Germany) with a UV and conductivity detector. The system was controlled by Dionex Chromeleon Version 7.1.2.1541. Separation of substances was performed via an anion-exchange column (6 µm, 2 × 250 mm, Dionex IonPac AS23, Thermo Scientific) with a guard column (0.25 mL min^{-1}, Dionex IonPac AG23, Thermo Scientific). An eluent concentrate (Dionex AS23, Thermo Scientific) was diluted to produce a solvent system of 4.5 mM sodium carbonate and 0.8 mM sodium bicarbonate. Nitrite as well as nitrate were detected at 210 nm. Identification and quantification was performed with reference substances (sodium nitrite and sodium nitrate, purity $\geq$ 99%, Carl Roth GmbH & Co. KG). The amount of hydrogen peroxide (H_2O_2) generated in the plasma-treated water was measured spectrophotometrically at 405 nm (UV-3100PC, VWR International, Hannover, Germany) after reaction with titanium sulfate.

2.3. Microorganisms and Culture Conditions

The bacteria *Staphylococcus aureus* DSM 799/ATCC 6538 (DSM-German Collection of Microorganisms and Cell Cultures, ATCC—American Type Culture Collection) and *Escherichia coli* K-12 DSM 11250/NCTC 10538 (NCTC—National Collection of Type Cultures) were cultured on tryptic soy agar plates (Carl Roth GmbH & Co. KG, Karlsruhe, Germany). After an incubation of 24 h at 37 °C the agar plates were stored at 8 °C.

For the experiments, the respective microorganism was cultured in 20 mL tryptic soy broth (Carl Roth GmbH & Co. KG, Karlsruhe, Germany). After an incubation time of 24 h at 37 °C, 10 mL of the culture was centrifuged (4700 rpm; Heraeus Multifuge 1S, Thermo Fisher Scientific) for 5 min. The supernatant was discarded and the cells were suspended in 10 mL physiological saline solution (0.85% NaCl). The bacteria suspension was adjusted to a total viable count of 10^6 cfu/mL (colony forming units/mL, stock suspension).

2.4. Determination of the Antimicrobial Effect of Plasma-Treated Liquids

The respective liquid-physiological saline solution or tap water was treated with plasma for 10, 20, and 30 min, defined as treatment time (t_{treat}). Afterwards, these liquids were mixed with the microorganism—*S. aureus* or *E. coli*—and incubated for 30 or 60 min, defined as exposure time (t_{exp}). For this purpose, 100 µL of the bacteria stock suspension was added to 5 mL of the plasma-treated liquid filled in a petri dish 55 × 14.2 mm (VWR, Darmstadt, Germany). The liquid was incubated on a multi-functional orbital shaker (PSU-20i, biosan, Riga, Latvia) for up to 30 or 60 min. Over the course of the exposure time, samples were taken in regular intervals and used to determine the total viable count as colony forming units/mL. The spiral plate method (spiral plater: Eddy Jet 2, IUL, Barcelona, Spain [29]) was used, with tryptic soy agar plates which were incubated for 24 h at 37 °C. The total viable count was measured by an automated colony counter (Flash & Go, IUL, Barcelona, Spain). As a control, 100 µL of the microorganism suspension was added to 5 mL of untreated liquid. This mixture was also used for the determination of the total viable count. All measurements were performed 6 times (n = 6).

The reduction of the total viable count is expressed as $\log_{10}(N_R)$ which was calculated as $\log_{10}(N_R)$ = $\log_{10}(N_0)$ − $\log_{10}(N_S)$, where N_0 is the total viable count of the control and N_S is the total viable count after the respective exposure time.

3. Results and Discussion

3.1. Electrical Characterization of the Discharge

To ignite a plasma between the metal electrodes and the surface of the liquid, we used a commercially available high voltage source usually used to operate gas discharge lamps [30]. This device provides two outputs with sinusoidal high voltages. As depicted in Figures 1 and 3, these voltages are phase-inverted to each other, meaning that the voltage on one electrode increases whereas the voltage on the other electrode decreases and vice versa. This leads to a sufficient potential difference between one electrode and the other with the liquid and the gas gaps in between. The discharge configuration could also work with grounded liquid, but this was not necessary in our configuration. Typical voltage and current waveforms for the two-electrode configuration with tap water as the liquid are shown in Figure 3. When the discharge occurs, the voltages at the electrodes (black and grey lines) collapse and the current (red and green lines) flows for some tens of nanoseconds. In these commercially available devices, high voltage is provided by use of leakage transformers to ensure short-circuit resistance. This means that the magnetic coupling between the primary and the secondary coil of the high-voltage transformer is weak, and thus, the output voltage collapses when the output is loaded and, consequently, the output current is limited. No additional current limiters like dielectric barriers or resistors are needed.

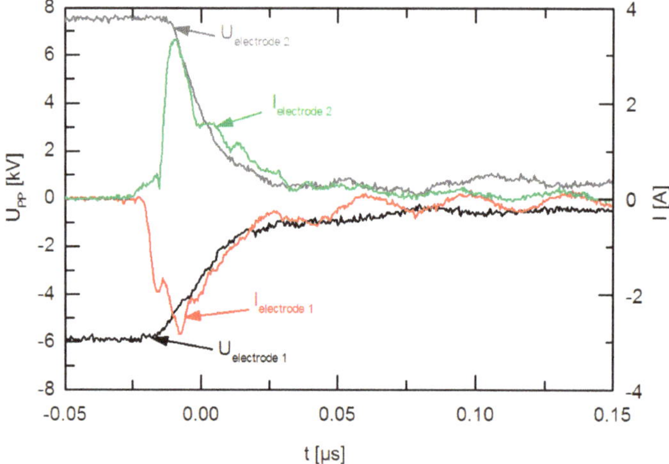

Figure 3. Electrical characterization of one discharge event with tap water as liquid [30]: Driving voltage at the electrodes (black and grey lines) and discharge current (red and green lines), discharges take place at both electrodes simultaneously.

In Figure 4, the influence of the liquid on the electrical parameters of the discharge is shown. For clearance, only the data for one electrode are presented. Generally, discharges occur very erratically, with varying breakdown voltage and different numbers of discharges per period. The presented waveforms are the most common. It was determined that for tap water (conductivity $\sigma = 0.648$ mS/cm) and saline solution ($\sigma = 16.23$ mS/cm) the voltage drop and the discharge current was significantly higher than for distilled water ($\sigma = 0.013$ mS/cm). The higher discharge current was the result of the higher initial conductivity of tap water and saline solution compared to distilled water (Figure 7). It also means a higher electrical load of the secondary coil of the output transformer, resulting in a higher magnetic stray flux, leading to a lower output voltage.

The significant influence of the conductivity of the liquid was also studied by Bruggeman et al. [31]. There, it was shown that a conductivity of 0.5–1 mS/cm leads to the highest current compared to larger and smaller conductivities. This was also confirmed in our experiments with the highest current using tap water ($\sigma = 0.648$ mS/cm).

Because of the erratic behavior of the discharges, the electrical power dissipated into the plasma was difficult to quantify. For a preliminary assessment of the efficiency of the process, the plug power (in this case the power dissipated into the high voltage source) was measured by a commercially available power meter. It was less than 50 W for the 4-electrode configuration, regardless of the treated liquid. This resulted in the consumption of less than 50 Wh for the production of 1 L plasma-treated water (equaling an expense of presently ≈ 0.02 € per liter for electricity and water).

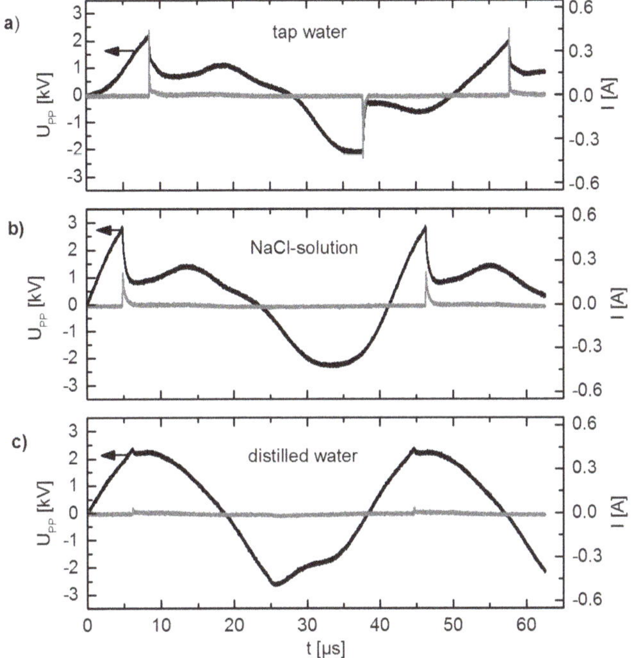

Figure 4. Electrical characterization for different treated liquids measured on one electrode: (**a**) tap water σ = 0.648 mS/cm, (**b**) saline solution σ = 16.23 mS/cm, (**c**) distilled water σ = 0.013 mS/cm.

3.2. Optical Characterization of the Discharge

It was visible with the naked eye that the color of the discharge was dependent on the treated liquid. The typical bluish/pinkish color of a non-thermal plasma operated in air changed to greenish when saline solution was treated instead of distilled water or tap water. The assumption that this was attributed to the emission of sodium was verified by optical emission spectroscopy (Figure 5). Significant emission of the Na-D-line at 589 nm (for the performed investigation it was not necessary to resolve the sodium doublet) was detected in the case of saline solution as the liquid. Although the nitrogen (N_2) emissions dominated the spectra, for tap water and saline solution, small emissions of nitrogen monoxide (NO) were also detected. Additionally, it was clearly shown that in the case of distilled water the general emission intensity was the lowest in good agreement with the lowest discharge current.

3.3. Chemical Characterization of the Liquids

Many researchers report strong acidification of the liquid by plasma treatment as long as non-buffered solutions are used [5,18]. The presence of a phosphate buffer inhibits the acidification [7,32]. Depending on the treated volume and the plasma source used, the acidification takes only one minute (V = 1.5 mL, [7]) or more than 10 min (V = 1 L, [5]). We treated in our experiments 500 mL of water for 30 min and opted for 10 min interval of measurement to investigate the general behavior of the liquid (Figures 6 and 7). In the slightly acidic (pH = 5.1) saline solution and demineralized water (pH = 4.8), the pH-value decreased to about 3 within 10 min treatment time whereas tap water (pH = 6.6) was not acidified at all. To our knowledge, this is due to the sufficient amount of buffering components like carbonate which corresponds to the high hardness of the local water used in this study (23.3° dH [33] compared to an average of 16° dH in Germany [34]).

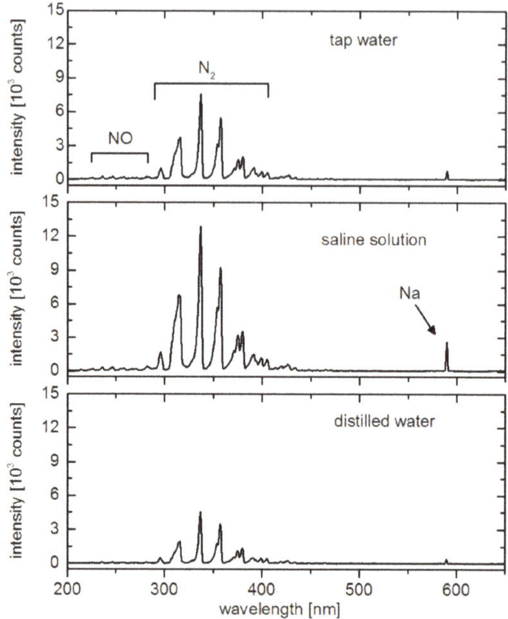

Figure 5. Optical characterization of the discharge for different liquids, the Na-D line is almost missing for liquids not containing sodium.

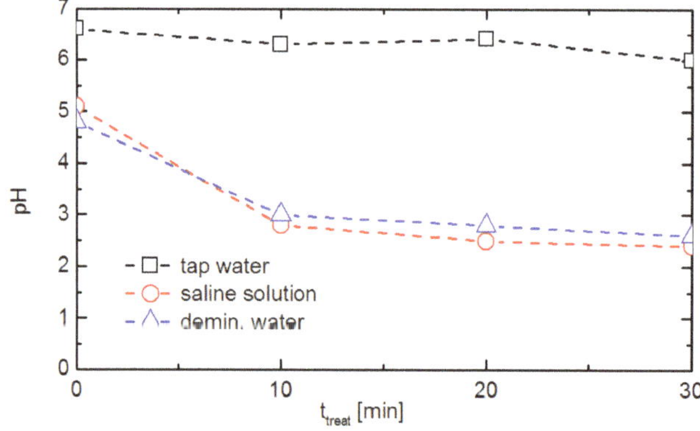

Figure 6. Change of pH-value with treatment time.

Figure 7 shows the change of the conductivity σ of the tested liquids. It was determined that the conductivity of the almost non-conductive distilled water increased significantly by around 300 µS/cm, which was in the range of the values reported in [4] ($\Delta\sigma$ = 150 µS/cm in 10 min). Also, for saline solution with a very high initial conductivity, an increase of about 700 µS/cm was measured. Again, we found different results for tap water. The increase in conductivity was less than 50 µS/cm after 30 min treatment.

After t_{treat} = 30 min, the concentrations of NO_2^- and NO_3^- were 1.55 mM and 0.77 mM, respectively (data not shown). Contrary to many other reported experiments, the concentration of H_2O_2 was negligible [6–8,32]. The optical emission spectra (Figure 5) show significant emissions of nitrogen and also emissions of nitrogen monoxide. This indicates that the nitration of the liquids is caused by the input of nitrogen species into the liquid by the discharge itself. Sivachandiran and Khacef [35] reported positive effects on germination and growth of plants such as radish, tomato, and sweet pepper with plasma treated water containing approximately 0.26 mM of NO_3^-. Thus, this kind of application could be an interesting field of research regarding a possible utilization of plasma treated water in agriculture. The temperature of the liquids increased slightly with a rate of $\Delta T \approx 0.7$ K/min. Therefore, also 30 min treatment does not cause notable evaporation of the liquid.

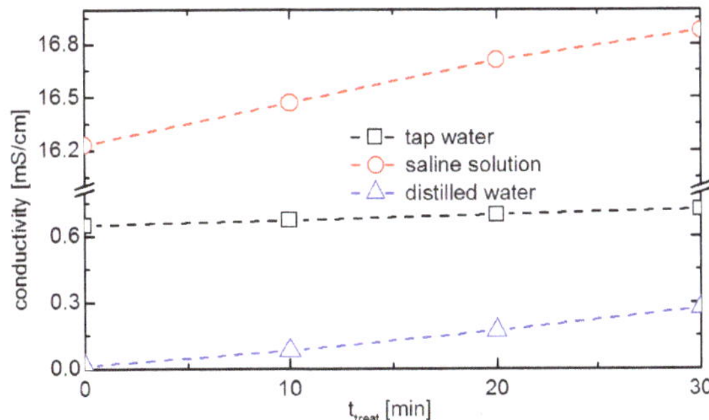

Figure 7. Change of conductivity with treatment time.

3.4. Antimicrobial Properties of the Plasma-Treated Liquids

In order to investigate the antimicrobial effect, a benchmark test was performed with plasma-treated physiological saline solution and the gram-negative bacterium *E. coli* K-12 DSM 11250/NCTC 10538. During the experimental procedure, the liquid was treated with plasma for 10, 20, or 30 min. The microorganisms were exposed to the plasma-treated liquid for up to 30 or 60 min (Figure 8; data for 60 min not shown).

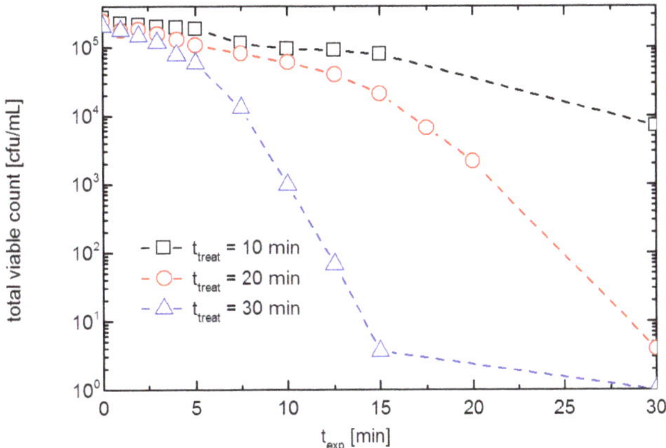

Figure 8. Antimicrobial effect of plasma-treated saline solution on the gram-negative bacterium *E. coli* for different treatment times: $t_{treat} = 10, 20$, and 30 min, determined by total viable count [cfu/mL] within an exposure time $t_{exp} = 30$ min.

The exposure of the microorganism *E. coli* to physiological saline solution treated with plasma for 10, 20 or 30 min resulted in a reduction of 5.3 orders of magnitude ($\log_{10}$ cfu/mL) within $t_{exp} = 30$–60 min. The inactivation of *E. coli* was dependent on the treatment time. Thus, within an exposure time of 10 min a reduction of total viable cell count of 0.5 orders of magnitude was determined for $t_{treat} = 10$ min whereas for $t_{treat} = 30$ min already a reduction of 2.3 orders of magnitude was reached. The fastest inactivation for *E. coli* occurred with the 30 min plasma-treated saline solution. For this

Article

Cold Argon Plasma as Adjuvant Tumour Therapy on Progressive Head and Neck Cancer: A Preclinical Study

Sybille Hasse [1,†,*], Christian Seebauer [2,†], Kristian Wende [1,3], Anke Schmidt [1], Hans-Robert Metelmann [2], Thomas von Woedtke [1,4] and Sander Bekeschus [1,3]

[1] Leibniz Institute for Plasma Science and Technology (INP), Felix-Hausdorff-Str. 2, 17489 Greifswald, Germany; kristian.wende@inp-greifswald.de (K.W.); anke.schmidt@inp-greifswald.de (A.S.); woedtke@inp-greifswald.de (T.v.W.); sander.bekeschus@inp-greifswald.de (S.B.)

[2] Department of Oral and Maxillofacial Surgery, Plastic Surgery, University Medicine Greifswald, Ferdinand-Sauerbruch-Str, 17475 Greifswald, Germany; seebauerc@uni-greifswald.de (C.S.); metelman@uni-greifswald.de (H.-R.M.)

[3] Leibniz Institute for Plasma Science and Technology (INP), ZIK Plasmatis, Felix-Hausdorff-Str. 2, 17489 Greifswald, Germany

[4] Department of Hygiene and Environmental Medicine, University Medicine Greifswald, Walther-Rathenau-Str. 49a, 17475 Greifswald, Germany

* Correspondence: sybille.hasse@inp-greifswald.de; Tel.: +49-3834-554-3921

† Those authors contributed equally to this paper.

Received: 29 April 2019; Accepted: 14 May 2019; Published: 19 May 2019

Abstract: Investigating cold argon plasma (CAP) for medical applications is a rapidly growing, innovative field of research. The controllable supply of reactive oxygen and nitrogen species through CAP has the potential for utilization in tumour treatment. Maxillofacial surgery is limited if tumours grow on vital structures such as the *arteria carotis*. Here CAP could be considered as an option for adjuvant intraoperative tumour therapy especially in the case of squamous cell carcinoma of the head and neck. Further preclinical research is necessary to investigate the efficacy of this technology for future clinical applications in cancer treatment. Initially, a variety of in vitro assays was performed on two cell lines that served as surrogate for the squamous cell carcinoma (SCC) and healthy tissue, respectively. Cell viability, motility and the activation of apoptosis in SCC cells (HNO97) was compared with those in normal HaCaT keratinocytes. In addition, induction of apoptosis in ex vivo CAP treated human tissue biopsies of patients with tumours of the head and neck was monitored and compared to healthy control tissue of the same patient. In response to CAP treatment, normal HaCaT keratinocytes differed significantly from their malignant counterpart HNO97 cells in cell motility only whereas cell viability remained similar. Moreover, CAP treatment of tumour tissue induced more apoptotic cells than in healthy tissue that was accompanied by elevated extracellular cytochrome c levels. This study promotes a future role of CAP as an adjuvant intraoperative tumour therapy option in the treatment of head and neck cancer. Moreover, patient-derived tissue explants complement in vitro examinations in a meaningful way to reflect an antitumoral role of CAP.

Keywords: cold argon plasma; head and neck squamous cell carcinoma; apoptosis; keratinocytes; plasma medicine

---

1. Introduction

Squamous cell carcinoma (SCC) is the second most frequent type of skin cancer after basal cell carcinoma [1] and arises frequently in the head and neck area. Treatment is multifaceted, challenging and has to be performed on an individualised basis. Complexity and variability of different tumour entities demand specific cancer therapies including surgery, chemotherapy or radiotherapy. Standard therapy for most of the extra- and intraoral tumours is based on a radical surgery combined with stage-based neo-adjuvant and / or adjuvant chemotherapy or radiation. Yet to date, patients suffering from advanced head and neck cancer are still faced with a poor prognosis, if cancer cells are infiltrating sensitive anatomical structures such as the skull base, the ocular bulb or main arteries that cannot be resected safely (one example is given in Figure 1). To reduce chemotherapy side effects and improve clinical outcomes, additional intraoperative antitumor treatment tools are urgently needed to complement current therapies. Cold atmospheric pressure plasma (CAP) at temperatures within physiological ranges offers versatile options for biological and medical applications in general [2–5] and recently was promoted to aid in cancer therapy [6–9].

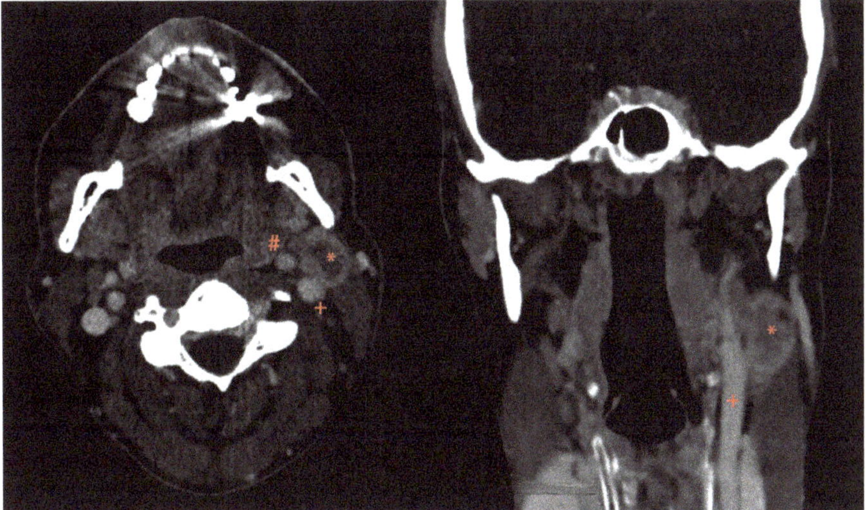

Figure 1. Axial (left) and coronal (right) cross sectional radiographic images of a lymph node metastasis (*) of the left cervical level IIb of a 38 years old male patient suffering from an advanced squamous cell carcinoma of the oral cavity at retromolar region (UICC stadium IVa). Primary carcinoma has been surgically removed with a surrounding safety margin. Due to the metastasis´ closeness to the vascular wall of the *arteria carotis interna* (#) and externa (+), a surgical removal could not be performed with a safety distance. Biopsy testing for the presence of possible remaining tumour cells at the vascular wall could not be performed. So, possible remaining tumour cells can be the reason for tumour recurrence or metastasis.

Plasma is often referred to as the fourth state of matter and is considered the most reactive one. Therapeutic effects are achieved in a single treatment modality due to the combined action of ultraviolet radiation, charged particles, reactive oxygen species (ROS), reactive nitrogen species (RNS) and electrical fields [2,10]. First results on the compatibility and application of CAP were gained in the field of dermatology as well as in plastic and aesthetic medicine [11–17]. CAP applications can range from anti-inflammatory [18] and antimicrobial [19,20] to tissue stimulatory [21] therapeutic effects. In addition, recently Vandamme et al. and others have shown that CAP can inhibit tumour growth by ROS-mediated apoptosis [6,22,23]. Moreover, there are indications that CAP works selective for cancer cells by killing them more effectively than non-neoplastic cells [24,25]. Consequently, the concept of using CAP in the manner of intra-operative neo-adjuvant tumour therapy is to treat remaining cancer cells after conventionally tumour bulk reduction by surgery as far as possible. Hence, CAP offers the potential for new adjuvant tumour therapies [6,8].

In this study, the response of healthy and tumour cells of head and neck cancer to CAP exposure was addressed by means of cell viability, cell cycle changes and motility behaviour, including possible target proteins and apoptosis. In addition, excised tissue samples of head and neck squamous cell carcinoma and healthy neighbouring tissue of the oral cavity of the same patient were investigated for induction of apoptosis and secreted proteins with antitumor activity. We were able to demonstrate a significant impact of CAP against malignant cells in tissue and partially in vitro. These insights are reason for hope that CAP could be an effective tool to support radical resection of head and neck cancers.

2. Methods

2.1. Cell Culture

Non-malignant HaCaT keratinocytes and malignant HNO97 (head and neck squamous cell carcinoma of the oral cavity) were employed in this study (Cell Line Service, Eppelheim, Germany). Both cell types were maintained in Roswell Park Memorial Institute (RPMI) 1640 cell culture medium supplemented with 8% heat inactivated foetal bovine serum, 100 IU/mL penicillin, 100 µg/mL streptomycin and 2 mM glutamine in a humidified atmosphere at 37 °C and 5% CO_2 (all PAN Biotech, Aidenbach, Germany).

2.2. Tissue Samples

Non-malignant and malignant tissue samples were collected from 10 patients (6 men, 4 women; mean age 59.9 years; median age 56 years; age range 43 to 75 years) with histologically proven cancers of the maxillofacial region [9 squamous cell carcinoma of the floor of the mouth (n = 5), the cheek (n = 2), the tongue (n = 1), the retromolar region (n = 1) and one adenocarcinoma of the maxillary sinus] during their preoperative inpatient care at the clinic of maxillofacial surgery / plastic surgery at the University Medicine Greifswald. Specimen were immediately immersed in William's medium (Biochrom, Cambridge, UK) supplemented with 0.25 µg/mL amphotericin B, 100 IU/mL penicillin and 10 µg/mL streptomycin (Corning, Amsterdam, The Netherlands). Subsequently, the tissue samples were treated with CAP and kept in serum free William's E medium supplemented with 100 IU/mL penicillin/ 10 µg/mL streptomycin, 2 mM glutamine, 10 µg/mL insulin and 10 ng/mL hydrocortisone (Sigma Aldrich, Taufkirchen, Germany) for 24 h, as described elsewhere [21,26]. All volunteers gave their written and informed consent to participate in the laboratory study and permitted the non-invasive collection of tissue samples for additional histological examinations after radical tumour surgery.

2.3. Plasma Source and Plasma Treatment

The certified medical device kINPen MED® (neoplas tools, Greifswald, Germany) was used to realize plasma treatment. It consists of two electrodes, a pin type high voltage electrode inside a ceramic capillary and one grounded electrode. It generates a radiofrequency signal of about 1 MHz and a voltage amplitude of 2–3 kV. The discharge is switched on and off at a frequency of 2.5 kHz (50:50). Argon gas (Air Liquide, Kornwestheim, Germany) flow rate was set to 5 standard litres per minute. The plasma was generated at the tip of the pin type electrode and expanded about 1 cm to the surrounding air outside the capillary.

In case of monolayer cells, the culture medium was exposed to the plasma effluent for the indicated length of time and plasma-treated medium was immediately added to the adherent cells whose cell culture medium was thereby replaced. By contrast, tissue samples were directly exposed to the plasma effluent at a distance of about 8 mm from the capillary outlet as previously described [27]. Untreated tissue samples served as negative control. The gas temperature at working distance is 35–39 °C.

2.4. Cell Viability

A test for cell viability was performed with a resazurin-based assay (*Alamar Blue®Assay*) and was evaluated as described in detail earlier [28]. Briefly, 3500 cells were seeded per well in 96-well plates and allowed to attach overnight. Plasma-treated cell culture medium (150 µL) was added to each well and incubated for 72 h. Plates were washed twice with Hanks buffered saline solution prior to the addition of 100 µM resazurin (Alfa Aesar, Kandel, Germany) in complete cell culture medium without phenol red. Formation of resorufin was observed at λ_{ex} 530 nm and at λ_{em} 590 nm using a microplate reader (Tecan, Männedorf, Switzerland). Cell viability was calculated as percentage of control (100%).

To investigate apoptosis in cell culture, *CellEvent* caspase 3/7 detection reagent (Life Technologies, Carlsbad, CA, USA) was added to cells 6 h after exposure to plasma-treated medium. Cell fluorescence was acquired at λ_{ex} 485 nm and λ_{em} 535 nm using the same microplate reader.

2.5. Cell Cycle Analyses

One million HaCaT and HNO97 cells, respectively, were seeded into 60 mm cell culture dishes. CAP-treated cell culture medium was added and incubated for 24 h. Cells were detached by incubation with trypsin/EDTA. Cells were transferred into FACS tubes, washed in PBS, fixed and permeabilised in ice-cold ethanol (70% v/v) overnight at 4 °C. Subsequently, cells were washed, and DNA was stained with 4,6-Diamidin-2-phenylindol (Sigma, Taufkirchen, Germany) in FACS-buffer for 1 h. DNA content analysis was carried out using flow cytometry (*Gallios*; Beckman-Coulter, Brea, CA, USA).

2.6. Cell Motility Assay

1×10^6 cells in 5 mL of fully supplemented RPMI 1640 were seeded in 60 mm cell culture dishes. Just before the cell culture medium was replaced by CAP treated medium a diagonal scratch was created using a 10 µL pipette tip. Dishes were placed on a heated and CO_2-gassed microscope stage (Zeiss, Jena, Germany). Microscopic images were taken at 24 h. The initial gap size was set to 0% and gap closure was calculated as percent.

2.7. DNA Fragmentation

As a marker for apoptosis in tissue, DNA fragmentation was detected by terminal dUTP nick-end labelling (TUNEL), employing a commercially available fluorescein kit (in situ cell death detection kit, Roche applied science, Mannheim, Germany). Positive cells were evaluated using fluorescence microscopy (Zeiss, Jena, Germany) and an open source image processing software (ImageJ, version 1.47, NIH, USA).

2.8. Cytochrome C Measurements

Twenty-four hours after CAP treatment, supernatant of cultured tissue samples were collected and stored at −80 °C until they were subjected to cytochrome c ELISA, according to the manufacturer's instructions (eBioscience, Affymetrix, ViennaAustria). Concentrations were normalized to total extracellular protein content (Biorad, Heracles, CA, USA).

2.9. Cytokine Detection

The same supernatant medium was analysed for tumour necrosis factor α (TNFα), interferon γ (INFγ), interleukin (IL) 10 and IL22 using a bead-based assay, according the manufacturer's instructions (BioLegend, San Diego, CA, USA). At least 300 beads per analyte were acquired utilizing a CytoFlex flow cytometer (Beckman-Coulter, Brea, CA, USA).

2.10. Global Protein Expression

Global protein expression was carried out as previously described [28]. Briefly, peptides were separated using nanoliquid chromatography (Dionex Ultimate 3000; PepMap RSLC column, 75 µm ID/15 cm length, Sunnyvale, CA, USA) and eluates were ionized by electrospray ionization and analysed by high resolution mass spectrometry (positive mode, QExactive, Thermo, Waltham, MA, USA). Data processing was done using Proteome Discoverer 1.4 (Thermo). Protein candidates were selected upon their involvement in pathways of metabolisms and proliferation as well as on statistical criteria (≥ ± 2.0-fold expression). Data were also analysed with Ingenuity Pathway Analysis software (IPA, Qiagen, Hilden, Germany) and free web-based applications (PANTHER and Uniprot.org).

2.11. Statistics

Statistical analysis was performed using *Prism* 6 (GraphPad software, version 6.05, San Diego, CA, USA). Two-way analysis of variances (ANOVA) with *Dunnett* correction (comparison to untreated control for each cell type separately) was employed to statistically describe cell motility data, whereas matched one-way ANOVA with *Dunnett* correction was applied to caspase, cell cycle, cytochrome c and cytokine data detection. The level of significance was displayed on the plots as follows: $p < 0.05$ (*), $p < 0.01$ (**), $p < 0.001$ (***).

3. Results

3.1. Examinations on Cultured Cells

3.1.1. CAP Reduced Viability in Normal and Tumour Cells

CAP expels reactive species which are known to confer cytotoxic effects in eukaryotic cells in a concentration dependent manner. HaCaT keratinocytes and HNO97 cells were exposed to plasma-treated medium and their metabolic activity was reduced in a treatment time-dependent manner (Figure 2A). In order to compare both cell lines an IC_{50} value was calculated by sigmoid regression of relative resorufin fluorescence as a function of treatment time. The viability after incubation in CAP treated medium between both cell lines revealed very similar IC_{50} values; that is, 31 s for HaCaT keratinocytes and 36 s for HNO97 cells (Figure 2A). This suggests a comparable sensitivity towards plasma-induced redox stress for these two cell types of epidermal origin.

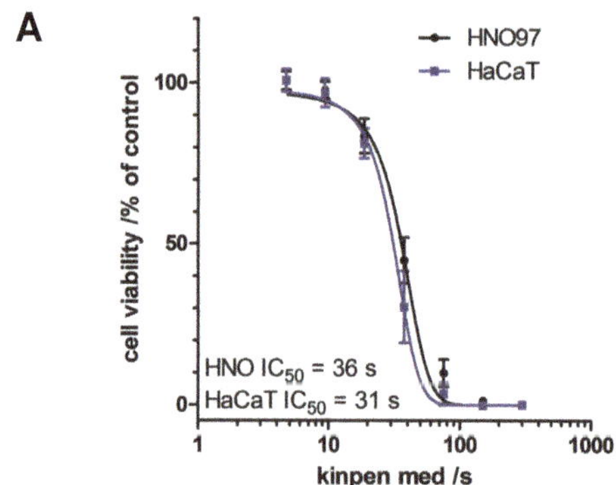

Figure 2. *Cont.*

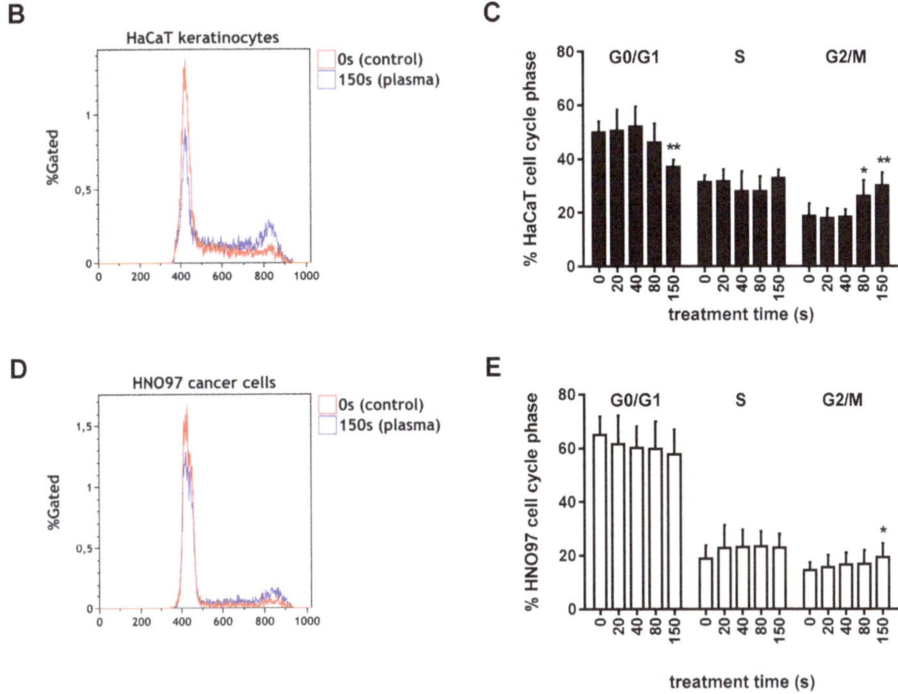

Figure 2. Comparison of the viability of HaCaT keratinocytes (blue line) and HNO97 (black line) in response to cold atmospheric pressure plasma (CAP) treated cell culture medium. (**A**). Cell viability is reduced after CAP treatment in both cell lines similarly. Formation of resorufin is plotted as a function of treatment time (log scale). Both curves summarize two individual experiments with 12 single measurements. 50% of viable cells (IC_{50}) are reached after CAP treatment for 31 s and 36 s, respectively. Cell cycle analyses revealed a G2/M-arrest in HaCaT keratinocytes and HNO97 cancer cells by CAP treated cell culture medium. Representative histograms for control (red line) and 150 s treatment time (blue line) are displayed for HaCaT keratinocytes (**B**) and for HNO97 cancer cells (**D**). Both, HaCaT keratinocytes and HNO97 cells responded by a G2/M arrest with increasing CAP exposure time, most pronounced at 80 s and 150 s (**C** and **E**).

3.1.2. Cell Cycle Arrest after CAP Treatment in HaCaT Keratinocytes and in HNO97 Cells

The distribution of DAPI-stained cells was analysed according to the histogram plot of DNA content against cell numbers of both cell lines (representative examples are shown in Figure 2B,D. In HaCaT keratinocytes, CAP treated medium (80 s and 150 s) caused a significantly elevated level of cells in the G2/M phase concomitant with a decrease in the proportion of cells in G0/G1 (Figure 2C). Of untreated cells 49% were in G0/G1 phase while this portion decreased significantly to 37% after 150 s CAP treatment. At the same time, cells in the G2/M phase increased from 18.6% to 30% for 150 s of CAP exposure. The impact of CAP treated medium on HNO97 cell cycle progression was less pronounced and revealed a decrease from 65% to 57.7% in the G0/G1 phase and a significant increase of 14.5% to 19.3% in the G2/M phase (Figure 2E).

3.1.3. CAP Exposure Activated Caspase in HNO97 Cancer Cells and HaCaT Keratinocytes

Both cell lines, normal and head and neck squamous cell carcinoma cells, significantly activated caspase 3/7 when incubated in CAP treated medium. (Figure 3). A treatment time dependent activation of caspase 3/7 was present after CAP exposure in both cell types. Hence, significant activation of caspase 3/7 levels occurred after the longest treatment times of 80 s and 150 s (Figure 3). This result indicates a non-selective induction of apoptotic events in both cell lines.

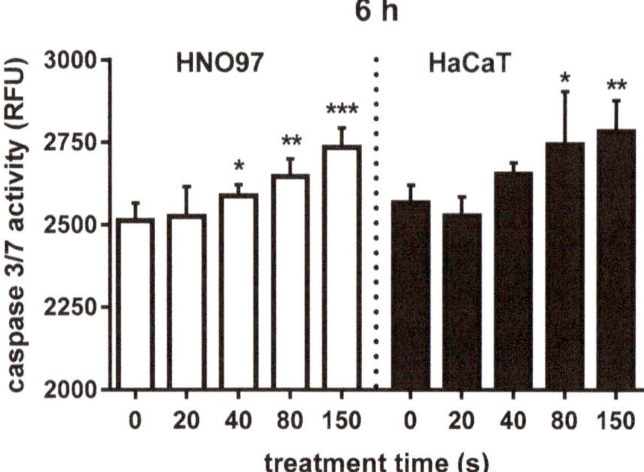

Figure 3. Activation of caspase 3/7 by CAP treated cell culture medium in HaCaT keratinocytes and HNO97 cells. CAP treatment took place for 0, 20, 40, 80 and 150 s and incubation followed for 6 h. Activation is detected as relative fluorescent units (RFU). Each bar resulted of three individual measurements (mean +SD).

3.1.4. Motility of Malignant Cells Was Significantly Impaired by CAP Treatment

Motility and migration were analysed in a 2D-setup following incubation with CAP-treated culture medium comparing non-malignant and head and neck squamous cell carcinoma cells. Into a confluent layer of cells, the scratch width was measured at start (0 h), after 12 h and 24 h following addition of CAP-treated culture medium. By comparison, untreated HNO97 cells moved significantly faster (Figure 4A) than untreated HaCaT keratinocytes (Figure 4B); gap closure within 24 h constituted 50% and 20%, respectively (Figure 4C). CAP treatment on HaCaT cells had no significant influence for the tested treatment time regimen (Figure 4C). However, in head and neck squamous cell carcinoma cells gap closure was significantly reduced from 50% in untreated cells to 20% after CAP treatment. In summary, CAP treatment caused a significant impairment in the motility behaviour of HNO97 while HaCaT keratinocytes remained nearly unaffected.

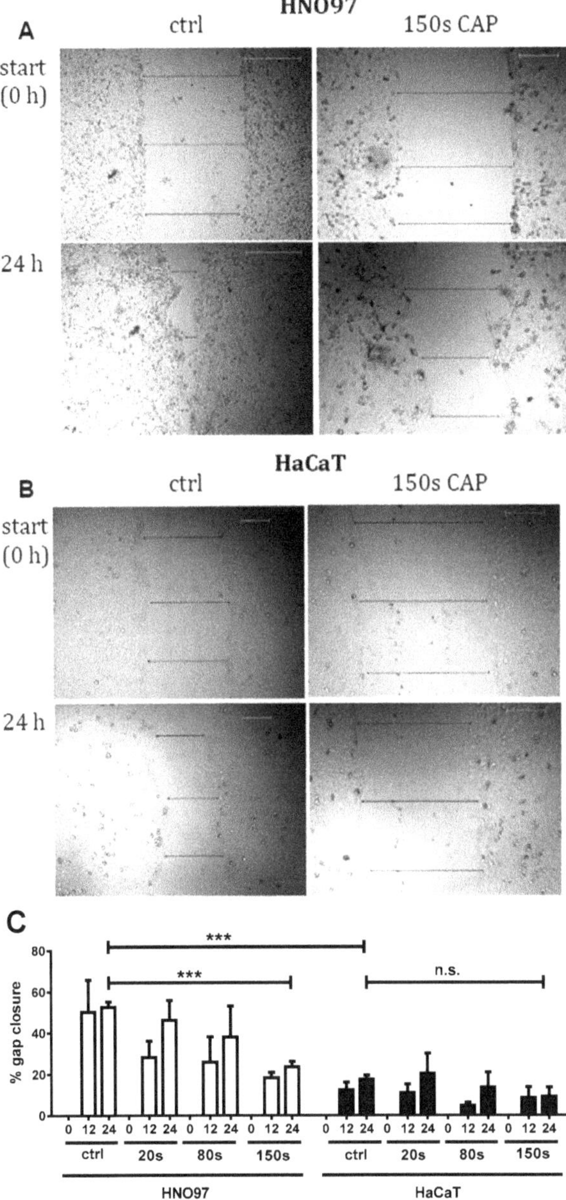

Figure 4. Cell motility was tested in HNO97 cancer cells and HaCaT keratinocytes in response to CAP treatment. Representative microscopic photographs of a scratch assay are depicted for (**A**) HNO97 and (**B**) HaCaT keratinocytes. Cell motility was evaluated as gap closure in % of the initial distance after 12 and 24 h, respectively and displayed as bar graph (**C**) for both cell lines. HaCaT keratinocytes are significantly less motile compared to their malignant counterpart HNO97 (compare both ctrl). Addition of CAP treated medium reduced cell motility significantly in HNO97 (empty bars) while HaCaT keratinocytes are not affected (dark bars). The assay was performed at least in triplicates. Bars represent mean + SD. Magnification × 100.

3.1.5. Total Protein Expression Modulated by CAP Treatment Reflect Changes in Cell Motility

Several thousand proteins were detected in the investigated cell lines with a false discovery rate of 1% (5445 HaCaT/8258 HNO97), allowing a materially oversight of cell processes at sampling. Of these, 229 HaCaT and 419 HNO97 proteins are connected to cell migration (according to Uniprot gene ontology annotations) with 185 proteins detected in both cell types. Selected proteins are summarized in the Table 1 below.

Table 1. Expression of selected proteins in normal and malignant cells 3 h and 24 h after CAP treatment. Protein changes are indicated as: up to two-fold increased →↑; two-fold increased ↑; 5 to 12-fold increased ↑↑; two-fold decreased ↓; 5-12-fold decreased ↓↓; ↔ not changed.

Protein Name	Uniprot ID	HaCaT	HNO97	Function
Ki67	P46013	↔	→↑	Proliferation
NQO1	P15559	→↑	↔	Nuclear factor erythroid 2-related factor 2 (NRF2) related signalling
TMX2	Q9Y320	↑	↔	NRF2 related signalling
GSTM3	P21266	↑	↔	NRF2 related signalling
PRDX1,2,4,6	Q06830	↔	↔	Redox regulation, elimination of peroxides
SBNO2	Q9Y2G9	↑↑	↑	Transcriptional co- regulation, interleukin signalling
ILKAP	Q9H0C8	↑↑	↔	Cell adhesion and growth factor signalling
AAMP	Q13685	→↑	↓	Cell migration, angiogenesis
ROCK2	O75116	→↑	↓↓	Regulation of actin cytoskeleton, cell adhesion and motility
CTTB2	Q8WZ74	↑↑	→↑	Cytoskeleton rearrangement, cell migration and motility
CFA20	Q9Y6A4	↑↑	↔	Cell motility
SRGP2	O75044	n.d.	↑	Cell migration inhibition, actin dynamics
ELMO3	Q96BJ8	↔	↓	Cell motility
CASP8/10	Q14790/Q92851	↔	↑	Pro-apoptotic caspases

3.2. Examinations on Tissue Specimen

Induction of apoptosis by CAP treatment was superior in tumour tissue compared with healthy mucosal tissue.

Following CAP triggered cytotoxicity and induction of early apoptotic events in cultured cells we next extended this view to biopsies of human non-malignant mucosa and squamous cell carcinoma of the head and neck area and investigated differences between these tissues after CAP treatment (3 min). The number of apoptotic cells was evaluated in situ in the tumour tissue and in healthy mucosal tissue of the same patient. In healthy tissue, DNA fragmentation after CAP exposure indicative for apoptosis was marginal. Low numbers of TUNEL-positive cells were counted, namely 2.58% versus 0.79% at median in untreated samples and CAP treated tissue specimens, respectively (Figure 5). In contrast, the tissue samples of head and neck squamous cell carcinoma displayed a significantly elevated median number of apoptotic cells of 5.6% compared to untreated tumour tissue (1.4%). In this context, it is noteworthy to mention that individual responsiveness varies (Figure 5).

To strengthen the finding of induced apoptosis by CAP in tissue samples, the supernatant medium was analysed for cytochrome c and selected cytokines. As a marker for cell death, cytochrome c is released by mitochondria upon induction of apoptosis. Initially, the tumour tissue even without CAP treatment revealed a much higher amount of cytochrome c than the corresponding healthy tissue (1729 ng per mg protein versus 385 ng per mg protein) (Figure 6). However, the CAP treatment triggered apoptotic pathways and the cytochrome c concentration raised significantly to as high as 5155 ng per mg protein in the supernatant of the tumour tissue (three-fold increase). Strikingly this was not the case for the corresponding healthy tissue. Here the same CAP treatment regimen led to a mean cytochrome c content of 698 ng per mg protein (1,8-fold increase) (Figure 6).

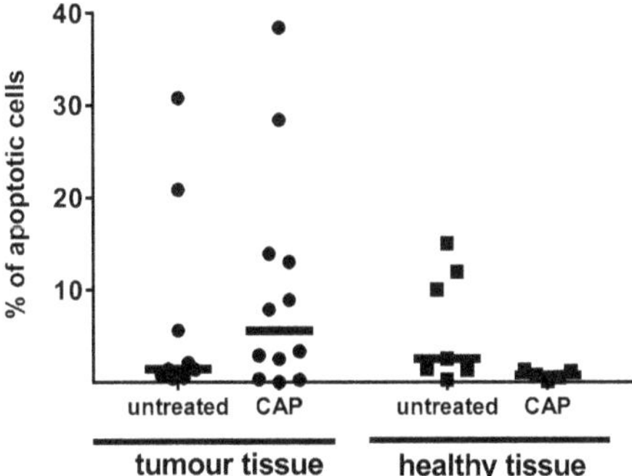

Figure 5. Induction of apoptotic cells by CAP application in tumour tissue in situ. TUNEL-positive cells were detected in several specimen of squamous cell carcinoma as well as in healthy tissue donated by the same patient. Results are displayed as scatter dot plot with a line at median. While healthy tissue revealed only very few TUNEL-positive cells after CAP exposure for 3 min the number increased in tumour tissue. Note the inter individual variance.

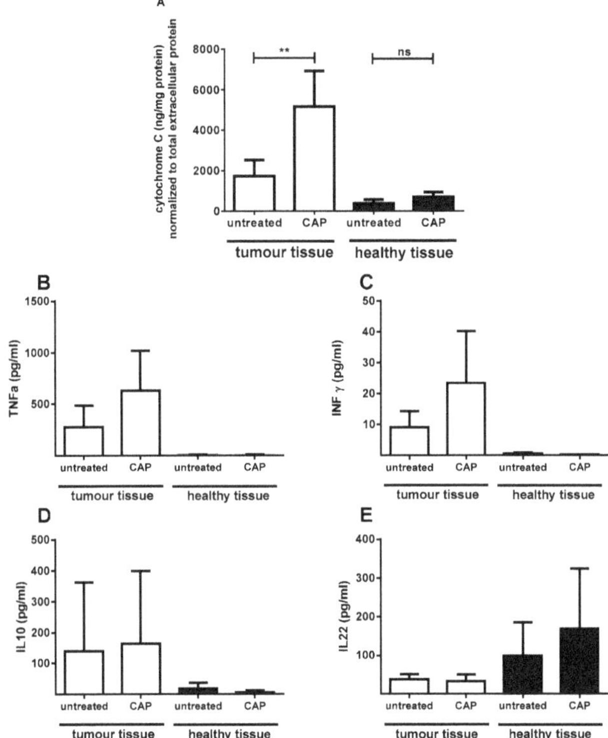

Figure 6. Detection of extracellular cytochrome c (**A**) and different cytokines (**B–E**) released by the tissue samples into the supernatant after CAP treatment. Secreted proteins were detected in the supernatant medium in which tissue specimen have been maintained for 24 h after CAP exposure. Higher extracellular cytochrome c levels were detected in tumour tissue compared to healthy mucosa of the same patient after CAP treatment (n = 9). The concentration of cytochrome c was normalized to total extracellular protein and bars represent mean + SEM. TNFα, IFNγ, IL10 and IL22 (n = 5) were simultaneously detected by a bead based immunoassay and compared between tumour tissue and healthy tissue. Bars represent mean + SEM.

Taken together, the results revealed that tissue of healthy and malignant origin differed significantly in their response to CAP treatment. Induction of apoptosis occurred predominantly in tumour tissue while in healthy tissue the impact of CAP was not significant. This finding inspired us to look for mediators like cytokines in conditioned medium in a preliminary test. Tumour necrosis factor α (TNFα), interferon γ (IFNγ), interleukin (IL) 10 and IL22 were detected and results are presented in Figure 6B–E. Due to small sample size and inter individual variances, only a tendency can be derived (Figure 6B–E). Strikingly higher concentrations of IL22 were released from healthy tissue while INFγ, TNFα and IL10 were found predominantly in tumour tissue of the corresponding patient (Figure 6B,C,D). CAP treatment tends to increase the levels of TNFα and IFNγ in tumour tissue but showed no effect in corresponding healthy tissue.

4. Discussion

The delivery of CAP-generated ROS and RNS through the liquid environment to cells changes the redox environment, which ultimately may induce cytotoxic effects [29,30]. With respect to cancer treatment, it is widely accepted that CAP-generated ROS and RNS are the main agents to induce cell death and apoptosis in cancer cells [6,24,31]. After CAP application an increasing number of apoptotic cancer cells positively correlate with an increased concentration of nitric oxide (NO), ROS and lipid peroxidation products [32]. Others demonstrated that CAP induces apoptosis via ROS and dysfunction of mitochondrial membranes [23]. In addition, autophagy has been discussed as another way of cell death after CAP exposure especially in primary cells [33]. This illustrates the great potential of CAP in the therapeutic concept of tumour treatment on the basis of ROS/RNS generation. In this context eliminating cancer cells by pharmacological ROS insults has also been developed independent of CAP technology [34]. The mode of action of some chemotherapeutic agents employs partly the production of ROS and therefore enhances their cytotoxic effectiveness (e.g., cisplatin or bleomycin) [35–38]. Even more so, a synergistic effect of gemcitabine and CAP in killing pancreatic tumour cells has been described recently [39].

Squamous cell carcinoma of the head and neck can grow in very close vicinity to vital structures (Figure 1), hence pose a surgical challenge for resection. Therefore, this study investigated CAP application as a potential adjuvant tool in the treatment of squamous cell carcinoma in this region. Unique advantages of the CAP application are the accessibility of localized areas and upon this time no known side effects [5,9].

In our study the certified medical device kINPen MED was employed to generate CAP. The therapeutic concept of the kINPen is based on the generation of ROS and RNS (RONS) [40]. Recently, a beneficial effect of this argon plasma jet treatment on head and neck cancer was documented in a retrospective clinical follow-up study by Metelmann and colleagues [41]. Several repeated cycles of CAP application for only one minute lead to partial remission of the tumour in 36% of the cases but prominent histological alterations were absent. Herein ex vivo treated tissue biopsies revealed an increased number of apoptotic cells within the tissue and elevated levels of cytochrome c in the extracellular liquid, pointing to profound apoptotic cell damage. This is in good agreement with clinical findings in head and neck squamous cell carcinoma patients [9]. In our approach, the distribution of TUNEL-positive cells was rather randomly dispersed throughout the tissue sections (data not shown), pointing to a non-cell-cell-contact-based mechanism. Contrary, results from Partecke et al. showed more dying cells in the outermost layers of treated pancreatic carcinoma tissue concomitant with reduced proliferation [42]. The therapeutic application of CAP in cancer treatment requires certain selectivity towards malignant cells. In our experimental setup we could show stronger induction of apoptosis in tumour tissue in situ compared to healthy tissue. In order to follow events in a 2D cell culture setup, two corresponding cell lines were selected, normal HaCaT keratinocytes and HNO97 derived from a human squamous cell carcinoma of the head and neck. However, the cell lines used revealed very little difference with respect to induction of apoptotic events after CAP exposure. Hence, the selectivity was not established in these in vitro tests. This may be due to the simple homogeneous cell population. Further the sensitivity towards CAP varies among HNSCC and non-malignant cell lines and the underlying molecular composition should be considered for future examinations. By contrast, others found that CAP selectively impairs some cell lines of head and neck squamous cell carcinoma also through non-apoptotic pathways [24,43]. The idea of CAP selectivity for cancer cells has been debated recently and a selective CAP effect on tumour cells has been ascribed to divergent metabolic features, anti-oxidative capacity or erroneous signalling processes [24,25,44,45].

In addition to cytotoxic events, the cells motility also was investigated. To preclude a risk, HaCaT and HNO97 cells were tested for their ability to move into a scratch. In this model, CAP treatment significantly decreased cell motility in squamous cell carcinoma cells only but not in non-malignant keratinocytes. Protein expression profiles support this finding by identifying reduced expression of several proteins related to cytoskeleton rearrangement, motility and cell migration. Cell motility as well as cell-cell-adhesion are both closely related to the expression of E-cadherin among other adhesion molecules. Previous research showed that the expression of E-cadherin negatively correlates with the progression of oral squamous cell carcinoma and dedifferentiation [46]. Interestingly, elevated E-cadherin and EGFR expression level were earlier detected after CAP treatment in HaCaT keratinocytes [47] but not in HNO97 cells.

Notably, the expression of angio-associated migratory cell protein (AAMP) also correlated with the migratory activity in several cell types, including smooth muscle cells and endothelial cells [48]. The expression of AAMP was reduced after CAP treatment in HNO97, whereas its expression slightly increased in HaCaT cells depending on the treatment intensity and time point [49]. We also observed that AAMP was secreted into the extracellular compartment and acts upon neighbouring cells, potentially contributing to remote effects of CAP.

Interestingly, for non-malignant murine fibroblasts a change in integrin expression concomitant with reduced motility after CAP exposure was reported earlier [50]. Moreover, we found that also SK-MEL-147 human melanoma cells displayed significantly decreased cell motility when subjected to CAP-treated medium [51]. This effect was correlated with alterations in the cytoskeleton, such as actin fibre rearrangement, and hence agrees well with alterations in non-melanoma skin cancer cells.

Regarding redox signalling events our results are in accordance with Weiss et al., who described unaltered protein levels of peroxiredoxin 1 and 2 after CAP exposure [52].

Moreover, our results illustrate the high complexity and heterogeneity of tumour tissue consisting of extracellular matrix, immune cells and epithelial cells. This led us to investigate extracellular cytokines after CAP exposure that function as mediators. Our results merely detected four immunomodulatory cytokines with different concentrations in healthy versus tumour tissue. Especially IL10 has been detected at higher levels in squamous cell carcinoma by others [53] and is in accordance with our finding. Considering the central role of INFγ in keratinocyte apoptosis and TNFα's potential to increase this effect, it is noticeable that our approach revealed an increased concentration of INFγ and TNFα after CAP treatment in tumour tissue [54]. This is in good agreement with detected elevated cytochrome c levels after CAP treatment. Effects mediated by second messenger like the AAMP protein or the interleukins and cellular signalling in response to CAP application remain largely elusive but require more attention in future approaches.

Intriguingly variations between HaCaT keratinocytes and HNO97 squamous cell carcinoma cells were rather minor in our 2D cell culture approach whereas patient derived tissue revealed very differential responses towards CAP treatment. This finding highlights the need to employ clinically relevant model systems for further preclinical investigations of CAP.

5. Conclusions

- Controlled application of CAP may provide a means to kill malignant cells.
- CAP application may be a promising adjuvant treatment option to eliminate minimal residual cancer cells after radical surgery of carcinoma of the head and neck area.

- Investigations towards the underlying mechanism remain to be addressed under consideration of accompanying cell types.

Author Contributions: Conceptualization, analysis, investigation: S.H., C.S., K.W., A.S., S.B.; writing—original draft preparation: S.H., C.S., A.S., S.B.; writing—review and editing, H.-R.M, T.v.W.

Funding: S.B. and K.W. received funding by the Federal Ministry of Education and Research (BMBF), grant no. 03Z22DN11 and 03Z22DN12.

Acknowledgments: The authors thank Liane Kantz for her excellent technical assistance.

Conflicts of Interest: The authors declare no conflict of interest.

Declarations: Ethics approval and consent to participate: Tissue sampling for ex vivo CAP treatment and investigation was approved by the ethics committee of the University Medicine Greifswald (approval number: BB61/11b). Informed consent was obtained from each patient prior to examinations.

References

1. Ferlay, J.; Steliarova-Foucher, E.; Lortet-Tieulent, J.; Rosso, S.; Coebergh, J.W.W.; Comber, H.; Forman, D.; Bray, F. Cancer incidence and mortality patterns in Europe: Estimates for 40 countries in 2012. *Eur. J. Cancer* **2013**, *49*, 1374–1403. [CrossRef]

2. Von Woedtke, T.; Reuter, S.; Masur, K.; Weltmann, K.D. Plasmas for medicine. *Phys. Rep.* **2013**, *530*, 291–320. [CrossRef]

3. Weltmann, K.D.; Kindel, E.; Brandenburg, R.; Meyer, C.; Bussiahn, R.; Wilke, C.; von Woedtke, T. Atmospheric Pressure Plasma Jet for Medical Therapy: Plasma Parameters and Risk Estimation. *Contrib. Plasma Phys.* **2009**, *49*, 631–640. [CrossRef]

4. Isbary, G.; Shimizu, T.; Li, Y.F.; Stolz, W.; Thomas, H.M.; Morfill, G.E.; Zimmermann, J.L. Cold atmospheric plasma devices for medical issues. *Expert Rev. Med. Dev.* **2013**, *10*, 367–377. [CrossRef]

5. Bekeschus, S.; Schmidt, A.; Weltmann, K.-D.; von Woedtke, T. The plasma jet kINPen—A powerful tool for wound healing. *Clin. Plasma Med.* **2016**, *4*, 19–28. [CrossRef]

6. Vandamme, M.; Robert, E.; Lerondel, S.; Sarron, V.; Ries, D.; Dozias, S.; Sobilo, J.; Gosset, D.; Kieda, C.; Legrain, B.; et al. ROS implication in a new antitumor strategy based on non-thermal plasma. *Int. J. Cancer* **2012**, *130*, 2185–2194. [CrossRef] [PubMed]

7. Keidar, M.; Shashurin, A.; Volotskova, O.; Ann Stepp, M.; Srinivasan, P.; Sandler, A.; Trink, B. Cold atmospheric plasma in cancer therapy. *Phys. Plasmas* **2013**, *20*, 057101. [CrossRef]

8. Schlegel, J.; Köritzer, J.; Boxhammer, V. Plasma in cancer treatment. *Clin. Plasma Med.* **2013**, *1*, 2–7. [CrossRef]

9. Schuster, M.; Seebauer, C.; Rutkowski, R.; Hauschild, A.; Podmelle, F.; Metelmann, C.; Metelmann, B.; von Woedtke, T.; Hasse, S.; Weltmann, K.D.; et al. Visible tumor surface response to physical plasma and apoptotic cell kill in head and neck cancer. *J. Cranio-Maxillo-Facial Surg.* **2016**, *44*, 1445–1452. [CrossRef]

10. Fridman, G.; Friedman, G.; Gutsol, A.; Shekhter, A.B.; Vasilets, V.N.; Fridman, A. Applied Plasma Medicine. *Plasma Process. Polym.* **2008**, *5*, 503–533. [CrossRef]

11. Heinlin, J.; Morfill, G.; Landthaler, M.; Stolz, W.; Isbary, G.; Zimmermann, J.L.; Shimizu, T.; Karrer, S. Plasma medicine: Possible applications in dermatology. *J. Dtsch. Dermatol. Ges.* **2010**, *8*, 968–976. [CrossRef] [PubMed]

12. Heinlin, J.; Schiffner-Rohe, J.; Schiffner, R.; Einsele-Kramer, B.; Landthaler, M.; Klein, A.; Zeman, F.; Stolz, W.; Karrer, S. A first prospective randomized controlled trial on the efficacy and safety of synchronous balneophototherapy vs. narrow-band UVB monotherapy for atopic dermatitis. *J. Eur. Acad. Dermatol. Venereol. JEADV* **2011**, *25*, 765–773. [CrossRef]

13. Isbary, G.; Heinlin, J.; Shimizu, T.; Zimmermann, J.L.; Morfill, G.; Schmidt, H.U.; Monetti, R.; Steffes, B.; Bunk, W.; Li, Y.; et al. Successful and safe use of 2 min cold atmospheric argon plasma in chronic wounds: Results of a randomized controlled trial. *Br. J. Dermatol.* **2012**, *167*, 404–410. [CrossRef]

14. Isbary, G.; Morfill, G.; Schmidt, H.U.; Georgi, M.; Ramrath, K.; Heinlin, J.; Karrer, S.; Landthaler, M.; Shimizu, T.; Steffes, B.; et al. A first prospective randomized controlled trial to decrease bacterial load using cold atmospheric argon plasma on chronic wounds in patients. *Br. J. Dermatol.* **2010**, *163*, 78–82. [CrossRef] [PubMed]

15. Metelmann, H.R.; Woedtke, T.; Bussiahn, R.; Weltmann, K.D.; Rieck, M.; Khalili, R.; Podmelle, F.; Waite, P. Experimental Recovery of CO_2-Laser skin lesions by plasma stimulation. *Am. J. Cosm. Surg.* **2012**, *29*, 52–56. [CrossRef]

16. Metelmann, H.-R.; Vu, T.T.; Do, H.T.; Le, T.N.B.; Hoang, T.H.A.; Phi, T.T.T.; Luong, T.M.L.; Doan, V.T.; Nguyen, T.T.H.; Nguyen, T.H.M.; et al. Scar formation of laser skin lesions after cold atmospheric pressure plasma (CAP) treatment: A clinical long term observation. *Clin. Plasma Med.* **2013**, *1*, 30–35. [CrossRef]

17. Brehmer, F.; Haenssle, H.A.; Daeschlein, G.; Ahmed, R.; Pfeiffer, S.; Gorlitz, A.; Simon, D.; Schon, M.P.; Wandke, D.; Emmert, S. Alleviation of chronic venous leg ulcers with a hand-held dielectric barrier discharge plasma generator (PlasmaDerm VU-2010): Results of a monocentric, two-armed, open, prospective, randomized and controlled trial (NCT01415622). *J. Eur. Acad. Dermatol. Venereol. JEADV* **2015**, *29*, 148–155. [CrossRef] [PubMed]

18. Nasruddin; Nakajima, Y.; Mukai, K.; Rahayu, H.S.E.; Nur, M.; Ishijima, T.; Enomoto, H.; Uesugi, Y.; Sugama, J.; Nakatani, T. Cold plasma on full-thickness cutaneous wound accelerates healing through promoting inflammation, re-epithelialization and wound contraction. *Clin. Plasma Med.* **2014**, *2*, 28–35. [CrossRef]

19. Daeschlein, G.; Scholz, S.; Ahmed, R.; von Woedtke, T.; Haase, H.; Niggemeier, M.; Kindel, E.; Brandenburg, R.; Weltmann, K.D.; Juenger, M. Skin decontamination by low-temperature atmospheric pressure plasma jet and dielectric barrier discharge plasma. *J. Hosp. Infect.* **2012**, *81*, 177–183. [CrossRef]

20. Daeschlein, G.; Scholz, S.; Arnold, A.; von Podewils, S.; Haase, H.; Emmert, S.; von Woedtke, T.; Weltmann, K.-D.; Jünger, M. In Vitro Susceptibility of Important Skin and Wound Pathogens Against Low Temperature Atmospheric Pressure Plasma Jet (APPJ) and Dielectric Barrier Discharge Plasma (DBD). *Plasma Process. Polym.* **2012**, *9*, 380–389. [CrossRef]

21. Hasse, S.; Duong Tran, T.; Hahn, O.; Kindler, S.; Metelmann, H.R.; von Woedtke, T.; Masur, K. Induction of proliferation of basal epidermal keratinocytes by cold atmospheric-pressure plasma. *Clin. Exp. Dermatol.* **2016**, *41*, 202–209. [CrossRef]

22. Sensenig, R.; Kalghatgi, S.; Cerchar, E.; Fridman, G.; Shereshevsky, A.; Torabi, B.; Arjunan, K.P.; Podolsky, E.; Fridman, A.; Friedman, G.; et al. Non-thermal plasma induces apoptosis in melanoma cells via production of intracellular reactive oxygen species. *Ann. Biomed. Eng.* **2011**, *39*, 674–687. [CrossRef] [PubMed]

23. Ahn, H.J.; Kim, K.I.; Hoan, N.N.; Kim, C.H.; Moon, E.; Choi, K.S.; Yang, S.S.; Lee, J.S. Targeting Cancer Cells with Reactive Oxygen and Nitrogen Species Generated by Atmospheric-Pressure Air Plasma. *PLoS ONE* **2014**, *9*, 1. [CrossRef]

24. Guerrero-Preston, R.; Ogawa, T.; Uemura, M.; Shumulinsky, G.; Valle, B.L.; Pirini, F.; Ravi, R.; Sidransky, D.; Keidar, M.; Trink, B. Cold atmospheric plasma treatment selectively targets head and neck squamous cell carcinoma cells. *Int. J. Mol. Med.* **2014**, *34*, 941–946. [CrossRef] [PubMed]

25. Keidar, M.; Walk, R.; Shashurin, A.; Srinivasan, P.; Sandler, A.; Dasgupta, S.; Ravi, R.; Guerrero-Preston, R.; Trink, B. Cold plasma selectivity and the possibility of a paradigm shift in cancer therapy. *Br. J. Cancer* **2011**, *105*, 1295–1301. [CrossRef]

26. Lu, Z.; Hasse, S.; Bodo, E.; Rose, C.; Funk, W.; Paus, R. Towards the development of a simplified long-term organ culture method for human scalp skin and its appendages under serum-free conditions. *Exp. Dermatol.* **2007**, *16*, 37–44. [CrossRef] [PubMed]

27. Hasse, S.; Hahn, O.; Kindler, S.; von Woedtke, T.; Metelmann, H.-R.; Masur, K. Atmospheric pressure plasma jet application on human oral mucosa modulates tissue regeneration. *Plasma Med.* **2014**, *4*, 431–438. [CrossRef]

28. Wende, K.; Reuter, S.; von Woedtke, T.; Weltmann, K.-D.; Masur, K. Redox-Based Assay for Assessment of Biological Impact of Plasma Treatment. *Plasma Process. Polym.* **2014**, *11*, 655–663. [CrossRef]

29. Graves, D.B. The emerging role of reactive oxygen and nitrogen species in redox biology and some implications for plasma applications to medicine and biology. *J. Phys. D Appl. Phys.* **2012**, *45*, 263001. [CrossRef]

30. Graves, D.B. Oxy-nitroso shielding burst model of cold atmospheric plasma therapeutics. *Clin. Plasma Med.* **2014**, *2*, 38–49. [CrossRef]

31. Graves, D.B. Reactive Species from Cold Atmospheric Plasma: Implications for Cancer Therapy. *Plasma Process. Polym.* **2014**, *11*, 1120–1127. [CrossRef]

32. Yan, K.; Kanazawa, S.; Ohkubo, T.; Nomoto, Y. Oxidation and reduction processes during NO(x) removal with corona-induced nonthermal plasma. *Plasma Chem. Plasma P* **1999**, *19*, 421–443. [CrossRef]

33. Hirst, A.M.; Simms, M.S.; Mann, V.M.; Maitland, N.J.; O'Connell, D.; Frame, F.M. Low-temperature plasma treatment induces DNA damage leading to necrotic cell death in primary prostate epithelial cells. *Br. J. Cancer* **2015**, *112*, 1536–1545. [CrossRef] [PubMed]

34. Trachootham, D.; Alexandre, J.; Huang, P. Targeting cancer cells by ROS-mediated mechanisms: A radical therapeutic approach? *Nat. Rev. Drug. Discov.* **2009**, *8*, 579–591. [CrossRef] [PubMed]

35. Engel, R.H.; Evens, A.M. Oxidative stress and apoptosis: A new treatment paradigm in cancer. *Front. Biosci.* **2006**, *11*, 300–312. [CrossRef]

36. Gorrini, C.; Harris, I.S.; Mak, T.W. Modulation of oxidative stress as an anticancer strategy. *Nat. Rev. Drug. Discov.* **2013**, *12*, 931–947. [CrossRef]

37. Hay, J.; Shahzeidi, S.; Laurent, G. Mechanisms of bleomycin-induced lung damage. *Arch. Toxicol.* **1991**, *65*, 81–94. [CrossRef] [PubMed]

38. Marullo, R.; Werner, E.; Degtyareva, N.; Moore, B.; Altavilla, G.; Ramalingam, S.S.; Doetsch, P.W. Cisplatin Induces a Mitochondrial-ROS Response That Contributes to Cytotoxicity Depending on Mitochondrial Redox Status and Bioenergetic Functions. *PLoS ONE* **2013**, *8*, 11. [CrossRef] [PubMed]

39. Masur, K.; von Behr, M.; Bekeschus, S.; Weltmann, K.-D.; Hackbarth, C.; Heidecke, C.-D.; von Bernstorff, W.; von Woedtke, T.; Partecke, L.I. Synergistic Inhibition of Tumor Cell Proliferation by Cold Plasma and Gemcitabine. *Plasma Process. Polym.* **2015**, *12*, 1377–1382. [CrossRef]

40. Reuter, S.; Tresp, H.; Wende, K.; Hammer, M.U.; Winter, J.; Masur, K.; Schmidt-Bleker, A.; Weltmann, K.D. From RONS to ROS: Tailoring Plasma Jet Treatment of Skin Cells. *IEEE Trans. Plasma Sci.* **2012**, *40*, 2986–2993. [CrossRef]

41. Metelmann, H.-R.; Nedrelow, D.S.; Seebauer, C.; Schuster, M.; von Woedtke, T.; Weltmann, K.-D.; Kindler, S.; Metelmann, P.H.; Finkelstein, S.E.; von Hoff, D.D.; et al. Head and neck cancer treatment and physical plasma. *Clin. Plasma Med.* **2015**, *3*, 17–23. [CrossRef]

42. Partecke, L.I.; Evert, K.; Haugk, J.; Doring, F.; Normann, L.; Diedrich, S.; Weiss, F.U.; Evert, M.; Hubner, N.O.; Gunther, C.; et al. Tissue Tolerable Plasma (TTP) induces apoptosis in pancreatic cancer cells in vitro and in vivo. *BMC Cancer* **2012**, *12*, 473. [CrossRef]

43. Welz, C.; Emmert, S.; Canis, M.; Becker, S.; Baumeister, P.; Shimizu, T.; Morfill, G.E.; Harreus, U.; Zimmermann, J.L. Cold Atmospheric Plasma: A Promising Complementary Therapy for Squamous Head and Neck Cancer. *PLoS ONE* **2015**, *10*, e0141827. [CrossRef] [PubMed]

44. Ma, Y.H.; Ha, C.S.; Hwang, S.W.; Lee, H.J.; Kim, G.C.; Lee, K.W.; Song, K. Non-Thermal Atmospheric Pressure Plasma Preferentially Induces Apoptosis in p53-Mutated Cancer Cells by Activating ROS Stress-Response Pathways. *PLoS ONE* **2014**, *9*, 4. [CrossRef] [PubMed]

45. Wang, M.; Holmes, B.; Cheng, X.; Zhu, W.; Keidar, M.; Zhang, L.G. Cold atmospheric plasma for selectively ablating metastatic breast cancer cells. *PLoS ONE* **2013**, *8*, e73741. [CrossRef] [PubMed]

46. Kaur, G.; Carnelio, S.; Rao, N.; Rao, L. Expression of E-cadherin in primary oral squamous cell carcinoma and metastatic lymph nodes: An immunohistochemical study. *Ind. J. Dent. Res.* **2009**, *20*, 71–76.

47. Haertel, B.; Wende, K.; von Woedtke, T.; Weltmann, K.D.; Lindequist, U. Non-thermal atmospheric-pressure plasma can influence cell adhesion molecules on HaCaT-keratinocytes. *Exp. Dermatol.* **2011**, *20*, 282–284. [CrossRef] [PubMed]

48. Beckner, M.E.; Jagannathan, S.; Peterson, V.A. Extracellular angio-associated migratory cell protein plays a positive role in angiogenesis and is regulated by astrocytes in coculture. *Microvasc. Res.* **2002**, *63*, 259–269. [CrossRef]

49. Schmidt, A.; Bekeschus, S.; Wende, K.; Vollmar, B.; von Woedtke, T. A cold plasma jet accelerates wound healing in a murine model of full-thickness skin wounds. *Exp. Dermatol.* **2017**, *26*, 156–162. [CrossRef]

50. Shashurin, A.; Stepp, M.A.; Hawley, T.S.; Pal-Ghosh, S.; Brieda, L.; Bronnikov, S.; Jurjus, R.A.; Keidar, M. Influence of Cold Plasma Atmospheric Jet on Surface Integrin Expression of Living Cells. *Plasma Process. Polym.* **2010**, *7*, 294–300. [CrossRef]

51. Schmidt, A.; Bekeschus, S.; von Woedtke, T.; Hasse, S. Cell migration and adhesion of a human melanoma cell line is decreased by cold plasma treatment. *Clin. Plasma Med.* **2015**, *3*, 24–31. [CrossRef]

52. Weiss, M.; Gumbel, D.; Hanschmann, E.M.; Mandelkow, R.; Gelbrich, N.; Zimmermann, U.; Walther, R.; Ekkernkamp, A.; Sckell, A.; Kramer, A.; et al. Cold Atmospheric Plasma Treatment Induces Anti-Proliferative Effects in Prostate Cancer Cells by Redox and Apoptotic Signaling Pathways. *PLoS ONE* **2015**, *10*, e0130350. [CrossRef] [PubMed]

53. Kim, J.; Modlin, R.L.; Moy, R.L.; Dubinett, S.M.; Mchugh, T.; Nickoloff, B.J.; Uyemura, K. Il-10 Production in Cutaneous Basal and Squamous-Cell Carcinomas—A Mechanism for Evading the Local T-Cell Immune-Response. *J. Immunol.* **1995**, *155*, 2240–2247.
54. Konur, A.; Schulz, U.; Eissner, G.; Andreesen, R.; Holler, E. Interferon (IFN)-γ is a main mediator of keratinocyte (HaCaT) apoptosis and contributes to autocrine IFN-γ and tumour necrosis factor-α production. *Br. J. Dermatol.* **2005**, *152*, 1134–1142. [CrossRef] [PubMed]

Article

Anticancer Efficacy of Long-Term Stored Plasma-Activated Medium

Ngoc Hoan Nguyen [1], Hyung Jun Park [2], Soon Young Hwang [1], Jong-Soo Lee [1,]* and Sang Sik Yang [2,]*

[1] Department of Life Sciences, Ajou University, Suwon 16499, Korea; hoanbiology@gmail.com (N.H.N.); hwang630@ajou.ac.kr (S.Y.H.)

[2] Department of Electrical and Computer Engineering, Ajou University, Suwon 16499, Korea; hjhj1201@ajou.ac.kr

* Correspondence: jsjlee@ajou.ac.kr (J.-S.L.); ssyang@ajou.ac.kr (S.S.Y.); Tel.: +82-31-219-1886 (J.-S.L.); +82-31-219-2481 (S.S.Y.)

Received: 21 November 2018; Accepted: 21 February 2019; Published: 25 February 2019

Abstract: The therapeutic potential of nonthermal atmospheric-pressure plasma for cancer treatment via generation of reactive species, induction of decreased mitochondrial membrane potential, and sequential apoptosis has been reported in our previous studies. Nonthermal atmospheric-pressure plasma-activated medium produced by jetting air plasma above a liquid surface shows advantages over direct plasma such as storage and delivery to tissues inside the body. In this study, we demonstrated that plasma-activated medium can be stored for up to 6 months in a freezer and that the stored plasma-activated medium has anticancer effects similar to those of direct plasma. Plasma-activated medium stored for 6 months showed cytocidal effects on human cervical cancer HeLa cells that were comparable to the effects of fresh plasma-activated medium or direct plasma. Furthermore, the levels of reactive species in plasma-activated medium persisted for up to 6 months. These results indicate that therapeutic application of plasma-activated medium is applicable in plasma medicine and is a promising anticancer strategy.

Keywords: plasma-activated medium; reactive oxygen species; apoptosis

---

1. Introduction

The fourth state of matter, plasma, is a partially ionized gas containing a high density of electrons and various reactive radicals and non-radicals. Recently, to enable the use of plasma in biomedical applications, various types of low-temperature atmospheric-pressure plasmas have been studied [1–6]. Dielectric barrier discharge is most often used to generate plasma because of its high electron density and insulation ability [4].

Cancer is among the leading causes of mortality worldwide and the treatment of patients with metastatic or relapsing cancer represents a main therapeutic challenge. New approaches to cancer therapy may involve the use of plasma. Biomedical applications of plasma have been demonstrated in a variety of experiments [5–11]. Recently, studies have revealed the ability of atmospheric-pressure plasma to selectively eliminate cancer cells, indicating the potential of nonthermal atmospheric-pressure plasma as an anticancer agent [1,3,7–10]. Nonthermal atmospheric-pressure plasma appears to be safe in the human body and normal cells but shows anticancer activity [1,3,7–10]. In contrast, plasma treatment may induce the death of normal primary prostate epithelial cells via DNA damage-mediated necrosis [11]. The selective and non-selective actions of plasma may occur through its dose-dependent differential effects on cancer and normal cells or cell type-dependent cytotoxic effects. Therefore, additional studies are needed to validate the potential therapeutic use of nonthermal atmospheric-pressure plasma, particularly against cancer.

We and many groups have reported various characteristics of atmospheric-pressure plasma, such as induction of apoptosis in cancer cells via DNA damage, mitochondrial collapse, and aberrations in the cellular membrane [1–3,5,9,12–19]. Additionally, recent studies revealed that reactive oxygen species (ROS) and reactive nitrogen species (RNS) generated by atmospheric-pressure plasma play crucial roles in inducing cancer cell death [1,3,10,15,16,19].

However, atmospheric-pressure plasma has two main limitations in biomedical applications. One is the limited penetration of plasma into tissues, the other is storage difficulties. To overcome these limitations, we employed plasma-activated medium (PAM) generated with nonthermal atmospheric-pressure plasma [3,8,20–25]. We evaluated the anticancer efficacy and storage feasibility of the PAM.

2. Materials and Methods

2.1. Reagents and Antibodies

We used the following reagents: fluorescein isothiocyanate (FITC) Annexin V Apoptosis Kit I (BD Pharmingen, Franklin Lakes, NJ, USA); Amplex UltraRed Kit (Life Technologies, Carlsbad, CA, USA); Griess Kit (Life Technologies); Live and dead cell assay Kit (Life Technologies); and MTT assay Kit (Abcam, Cambridge, UK).

2.2. Micro Plasma-Jet Nozzle

The structure of the micro plasma-jet nozzle for the nonthermal atmospheric-pressure plasma jet is shown in Figure 1a. The nozzle consists of an anode, stainless-steel electrode ring, and cathode. The anode was a 100-μm-thick nickel-cobalt layer on a 500-μm-thick glass wafer. The anode contained 25 holes with a diameter of 300 μm. The pattern of the hole arrangement was described previously [26]. The cathode was a stainless-steel tube through which gas was supplied and the inner diameter of the tube was 1.5 mm. The anode was fabricated by photolithography and nickel-cobalt electroplating with a Cr/Au seed layer on a glass wafer. After chemical mechanical polishing of the glass wafer, the anode holes were fabricated by sand blasting. These electrodes were packaged in a plastic case.

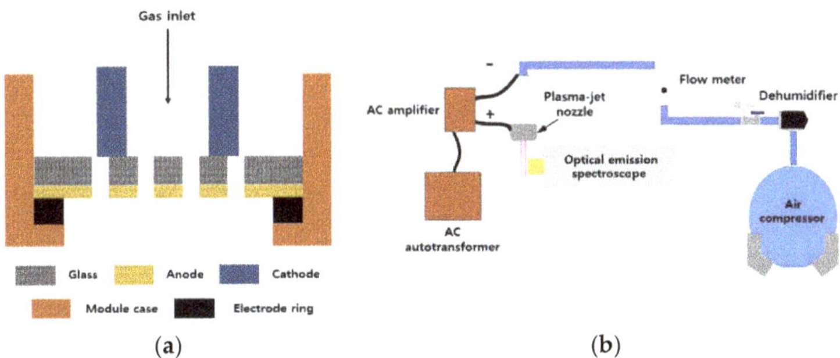

Figure 1. Nonthermal atmospheric-pressure plasma-jet system. (**a**) Structure of the micro plasma-jet nozzle. (**b**) Experimental set-up for discharge experiments.

2.3. Nonthermal Atmospheric-Pressure Plasma-Jet System

Figure 1b shows the experimental set-up for the discharge experiments using the nonthermal atmospheric-pressure plasma-jet system. The power controller was a variable AC autotransformer supplying 15 $kV_{p\text{-}p}$ at a frequency of 20 kHz. Air was supplied through the tube at 10 L/min. We recorded the optical emission spectrum of the plasma jet using an optical emission spectroscope (SV 2100, K-MAC, Daejeon, Korea, 2012) to identify radicals useful for biomedical applications.

2.4. Preparation of Plasma-Activated Medium (PAM)

As shown in Figure 2, using this air plasma-jet system, we jetted nonthermal air plasma 2 cm above the surface of Dulbecco's Modified Eagle Medium (DMEM) (WELGENE, Daejeon, Korea) supplemented with 10% fetal bovine serum (FBS) and antibiotics (Life Technologies), a mammalian cell culture medium, in a chamber of a 12-well-plate for 5 min at atmospheric pressure and room temperature to generate PAM. The height and diameter of the well chamber were 2.5 and 2.2 cm, respectively. We evaluated the dissolved ozone concentration in the PAM using a waterproof portable colorimeter (C105, EUTECH, Singapore). For the storage test, we stored the PAM in a freezer ($-20\,^{\circ}$C) for up to 6 months.

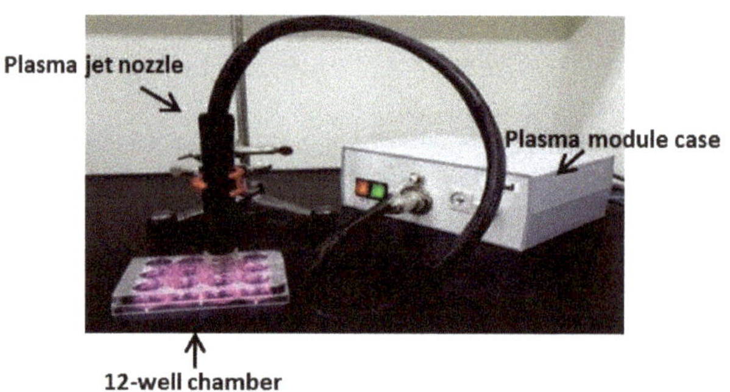

Figure 2. Generation of plasma-activated medium (PAM) using DMEM by jetting air plasma 2 cm above the surface of the medium in a well (2.5×2.2 cm) of a 12-well-plate.

2.5. Quantification of Reactive Oxygen Species (ROS) and Reactive Nitrogen Species (RNS)

The concentrations of extracellular H_2O_2 were measured using an Amplex UltraRed hydrogen peroxide assay kit (Invitrogen, Carlsbad, CA, USA). Florescence intensity was measured on a microplate reader (Bio-Rad, Hercules, CA, USA) at 530/590 nm according to the manufacturer's protocol. The relative ROS level was calculated in terms of arbitrary fluorescence units. The production of RNS in the culture supernatant by plasma was detected as described previously using a Griess assay kit (Invitrogen).

2.6. Cell Culture and Treatment with PAM

Human cervical carcinoma HeLa cells were obtained from the American Type Culture Collection (Manassas, VA, USA) and cultured in DMEM supplemented with 10% FBS and antibiotics. All cells were maintained at 37 $^{\circ}$C in a humidified incubator at 5% CO_2. Cells (approximately 5×10^4) were seeded into 12-well plates with 1 mL of DMEM supplemented with 10% of FBS and grown overnight before treatment with the fresh/stock PAM. The cells were washed with Dulbecco's phosphate-buffered saline (Life Technologies) and covered with either the fresh/stock gas-treated DMEM or fresh/stock PAM. Next, the cells were incubated at 37 $^{\circ}$C for 24 h to assess the anticancer efficacy of the PAM.

2.7. Detection of Apoptosis

During treatment with PAM, the cells were incubated at 37 $^{\circ}$C in a humidified incubator at 5% CO_2 for 24 h. The cells were harvested with trypsin-EDTA and rinsed with phosphate-buffered saline. To detect plasma-induced apoptosis, we stained the cells with FITC-conjugated anti-annexin V or propidium iodide (Invitrogen). After staining, the cells were analyzed by flow cytometry (BD FACSAria III).

2.8. Measurement of Cellular Viability

After 24 h of treatment with either the fresh/stock PAM or direct plasma, the live/dead assay involved double-labeling of the cells with 2 µM calcein AM and 4 µM EthD-1 according to the manufacturer's instructions (Life Technologies and Abcam). Calcein AM-positive live cells and EthD-1-positive dead cells were examined under a fluorescence microscope (Nikon Inverted Microscope Eclipse Ti-S/L100, Tokyo, Japan). Additionally, a 3-(4,5-dimethylthiazol-2-yl)-2,5-diphenyltetrazolium bromide (MTT) colorimetric assay (Abcam) was performed to evaluate cellular viability and activity as previously described [1,19].

2.9. Statistical Analysis

All data were expressed as mean $\pm$ standard deviation (SD) of at least three replicates. Student's *t* test was used for the analysis of significance of differences between datasets. Differences were considered statistically significant at $p \leq 0.05$ (in figures: * $p \leq 0.05$, ** $p \leq 0.01$, *** $p \leq 0.001$).

3. Results

3.1. Reactive Species Generated in Plasma Jet and PAM

Figure 3a shows the optical emission spectrum of the nonthermal atmospheric-pressure air plasma jet in the atmosphere 1 cm apart from the nozzle end over a wide range of wavelengths from 280 to 920 nm. The air plasma contained excited oxygen ions (O_2+) and excited nitrogen molecules as the ROS and RNS, respectively.

Dissolution of the plasma indicates that the plasma-jet system is useful for various biomedical applications [27]. We adopted dissolved ozone among various reactive oxygen species in order to find an appropriate plasma-treatment time. Figure 3b shows that the concentration of dissolved ozone in plasma-treated DMEM increases with increasing plasma-treatment time. Since the concentration reached a saturation level of 0.3 ppm in 5 min, the plasma-treatment time was set to 5 min in the preparation of PAM.

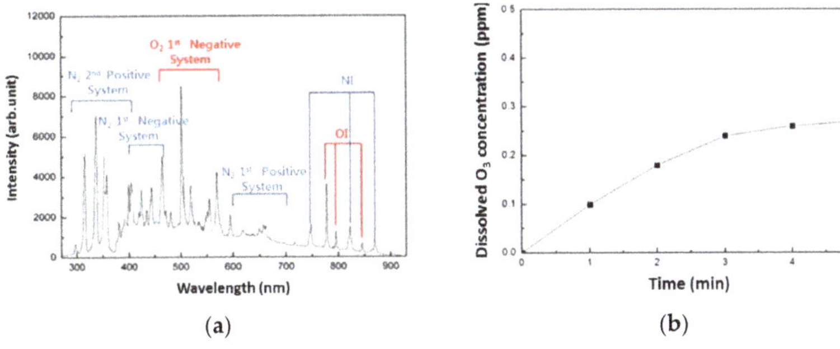

Figure 3. Reactive species generated in plasma jet and PAM. (**a**) Optical emission spectrum of air plasma jet. (**b**) Concentration of dissolved ozone in Dulbecco's Modified Eagle Medium (DMEM) vs. plasma treatment time.

3.2. Anticancer Efficacy of PAM

Recent studies revealed that treatment of cancer cells with direct nonthermal plasma induces apoptotic cell death, accompanied by ROS accumulation, reduced mitochondrial membrane potential, and mitochondrial dysfunction [1,19]. To use of plasma in medical applications, there are several essential prerequisites including long-term storage and efficient delivery to internal parts of the human body. Therefore, we first evaluated the anticancer effects of PAM compared to those of direct plasma. After 24 h of incubating the cells with PAM or gas-treated medium or after 24 h of incubating the cells exposed to direct plasma for 5 min, cell viability assay was conducted by double-labeling of the cells with calcein AM and EthD-1. Calcein AM-positive live cells and EthD-1-positive dead cells were assessed under a fluorescence microscope (Figure 4a). Additionally, cell viability and activity were evaluated by MTT assay after plasma treatment (Figure 4b). PAM caused the death of human cervical cancer HeLa cells, analogously to direct air plasma, while gas treatment had no effects on cancer cell viability (Figure 4). This suggests that PAM has comparable anticancer effects as direct plasma (Figure 4).

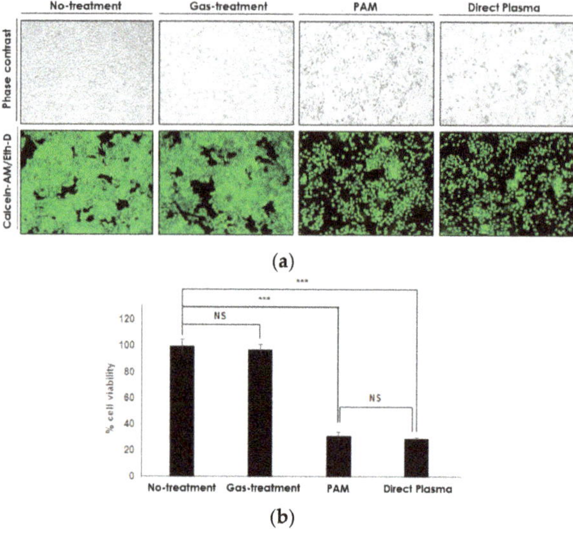

Figure 4. Comparison of anticancer effects of PAM and direct plasma. (**a**) Human cervical HeLa cancer cells were treated with either PAM or direct plasma, or gas (gas-treatment), stained with calcein-AM and EthD-1, and their viability was examined under a fluorescence microscope. (**b**) Cell viability and activity were assessed by MTT assay and compared to those of untreated or gas-treated control cells. NS, not significant; * $p \leq 0.05$, ** $p \leq 0.01$, *** $p \leq 0.001$.

3.3. Anticancer Efficacy of Frozen PAM

Next, we evaluated the period of efficacy of PAM. PAM was stored frozen and thawed at the indicated time-points to treat HeLa cancer cells. The results in Figure 5a–b show no significant differences in the effects on HeLa cancer cell viability between fresh and stocked PAM that had been frozen and thawed. The effect was also comparable to that observed for direct plasma treatment. Additionally, we examined whether fresh and stock PAM induced apoptosis in cancer cells after 24 h of treatment with PAM by staining the cells with annexin V conjugated with FITC and PI. After staining, cell death was analyzed by flow cytometry. Apoptosis was detected by labeling phosphatidylserine on the membrane surface of apoptotic cells with annexin V, and late apoptosis/necrosis were detected by labeling cellular DNA with PI. Figure 5c shows that 1- and 6-month stock PAM caused similar levels of cancer cell death and most PAM-treated cancer cells underwent apoptosis.

Further, we quantitatively assessed the anticancer effects of fresh and stock PAM by diluting PAM and then evaluating its cytotoxic effects on HeLa cells in a live/dead cell assay (Figure 6a) and MTT assay (Figure 6b). Dilutions of PAM by half or 1/4 efficiently induced cell death of HeLa cells but caused slightly lower cell death as compared to undiluted fresh and stock PAM (Figure 6). However, 1/10-diluted PAM did not cause cell death of HeLa cells. Thus, we estimated that the anticancer effect of fresh and stock PAM ranged from 1/10 to 1/4 dilutions. Moreover, we observed a similar dilution effect between the fresh and stock PAM (Figure 6), indicating that the components required for the anticancer effects in the stock PAM do not decline over time.

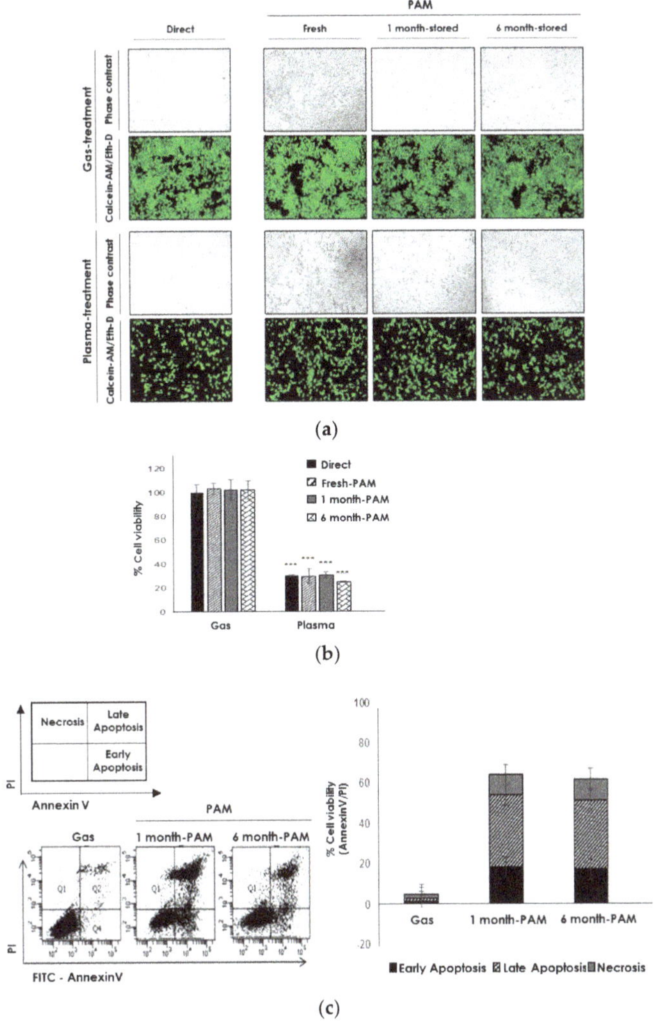

Figure 5. Efficacy of frozen PAM. PAM was stored in a freezer ($-20\ ^{\circ}$C) for the indicated periods (1 and 6 months), then thawed and added to HeLa cells to evaluate its efficacy. (**a**) Human cervical HeLa cancer cells were treated with plasma (direct plasma or fresh and 1- or 6-month stock PAM) (plasma-treatment) or gas (gas or gas-treated fresh and stock medium) (gas-treatment) for 24 h. Cellular viability was assessed using calcein-AM and EthD-1 to detect live and dead cells, respectively. (**b**) Cell viability was measured by MTT assay and compared to those of gas-treated control cells. (**c**) Cell death via apoptosis/necrosis was detected by staining the cells with FITC-conjugated annexin V and propidium iodide (PI), respectively, and by flow cytometric analysis, and compared to that in gas-treated cells. The dot plots show representative images of apoptotic cell death after PAM-treatment, identified by labeling phosphatidylserine on the membrane surface of early/late apoptotic cells with annexin V, and by labeling DNA of late apoptotic/necrotic cells with PI. The graph shows the populations of apoptosis/necrotic cells.

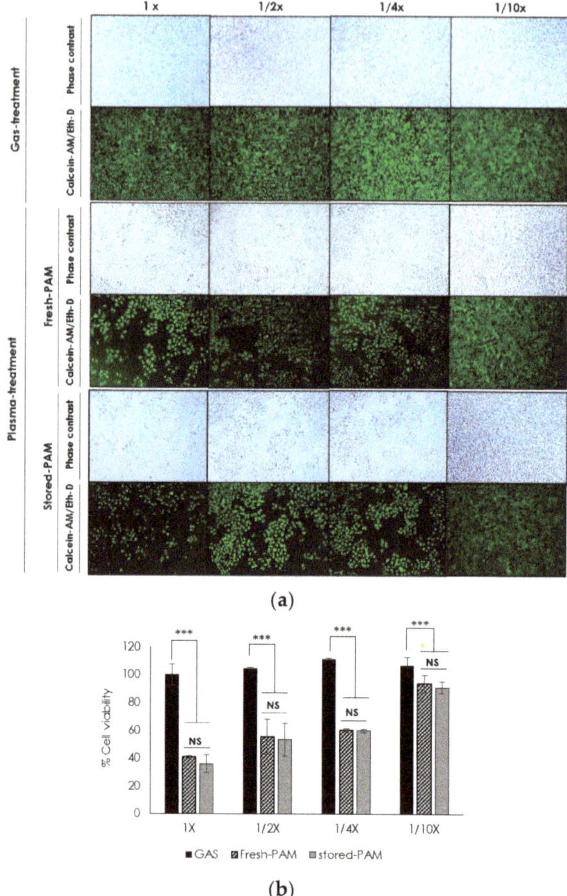

(a)

(b)

Figure 6. Quantitative assessment of anticancer efficacy of stock PAM. The fresh and 6-month stock PAM (1×) were diluted by 1/2, 1/4, and 1/10. HeLa cells were incubated in the diluted fresh and stock PAM for 24 h and then cell viability was evaluated by conducting live/dead assay (**a**) and MTT assay (**b**). NS: not significant; * $p \leq 0.05$, ** $p \leq 0.01$, *** $p \leq 0.001$.

3.4. Maintained ROS Level in Stock Plasma Indicating the Advantage of PAM

We then monitored the levels of ROS and RNS in fresh and stock PAM (Figure 6). In agreement with the cell death levels induced by fresh and stock PAM (Figures 4–6), the levels of ROS and RNS in fresh and stock PAM were similar (Figure 7). We performed a hydrogen peroxide assay and Griess assay to analyze ROS (H_2O_2) and RNS (nitrite and nitrate), respectively. Interestingly, significantly increased levels of H_2O_2 were detected in fresh and stock PAM (37.2–40.5 µM) (Figure 7a), suggesting that the level of H_2O_2 in stock PAM was stably maintained. Modest increases in nitrite and nitrate measured by the Griess assay were observed in both fresh and stock PAM (Figure 7b) compared to in control medium, and then gradually decreased to the basal level observed in control gas-treated medium, indicating that nitrite and nitrate were sustained in the stock PAM. Taken together, these results demonstrate that PAM can be effectively stored for up to 6 months, possibly without a significant decrease in the levels of reactive ingredients.

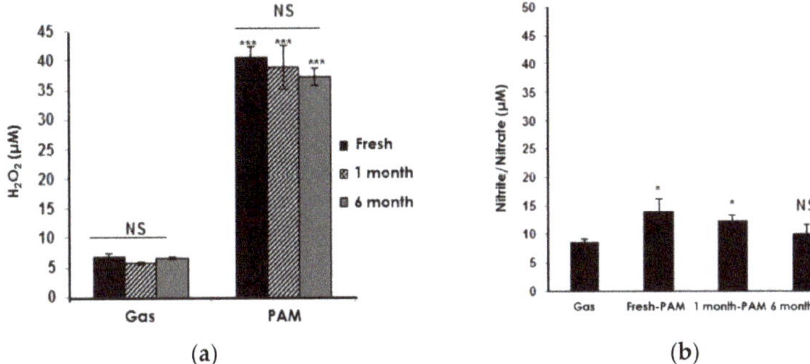

Figure 7. Reactive species in stock PAM were stably maintained. H_2O_2 (**a**) and nitrite/nitrate (**b**) in fresh and stock PAM were quantified. PAM was stored in a freezer ($-20\ ^\circ$C) for the indicated periods (1 or 6 months) and then thawed to assess H_2O_2 and nitrite/nitrate levels. DMEM served as a control. NS, not significant; * $p \leq 0.05$, ** $p \leq 0.01$, *** $p \leq 0.001$.

4. Discussion

Consistent with the results of our previous study, we demonstrated that PAM had anticancer effects on HeLa cells [1,3,28]. PAM had the same effect on inducing cancer cell death as direct treatment.

The rates of cell viability or apoptosis after treatment with fresh or stock PAM were detected by live/dead, flow cytometry, or MTT assays. No large differences were observed between the fresh and stock PAM. Both plasmas induced cell death via apoptosis. This suggests that fresh and stock PAM induce cell apoptosis through the same mechanism. Our data showed that the concentration of dissolved ozone, a reactive oxygen species (ROS), increased with increased exposure time. Along with the optical emission spectrum data, the level of ROS was increased by 10-fold in treated culture media compared to non-treatment; this rate was similar in fresh and stock PAM. Additionally, RNS (nitrite and nitrate) in treated culture media showed a modest increase compared to non-treatment and a gradual decrease during stock. Consistent with previous reports [21,22,28], our findings indicate that the anticancer effect of fresh and stock PAM depend on H_2O_2 and nitrite/nitrate. Our study suggests that fresh and frozen PAM function as apoptotic inducers by generating ROS/RNS products in cellular cultured media.

Based on our results, PAM may be useful as a pharmaceutical drug. PAM shows anticancer effects equivalent to those of direct air plasma and the stock PAM showed anticancer efficacy. In summary, our results broaden the application spectrum and enhance the feasibility of plasma medicine.

Author Contributions: Authors' individual contributions are as follows: plasma generation, H.J.P.; biological experiments, N.H.N. and S.Y.H.; writing—original draft preparation, N.H.N. and H.J.P.; writing—review, J.-S.L.; writing—editing, S.S.Y.; funding acquisition, J.-S.L. and S.S.Y.

Funding: This research was supported by National Research Foundation of Korea (NRF) grants funded by the Ministry of Science, ICT, & Future Planning (grant numbers 2012M3A9B2052871, 2012M3A9B2052872, 2014R1A2A2A01004444, and 2011-0030043).

Conflicts of Interest: The authors declare no conflict of interest. The funders had no role in the writing of the manuscript.

References

1. Ahn, H.J.; Kim, K.I.; Hoan, N.N.; Kim, C.H.; Moon, E.; Choi, K.S.; Yang, S.S.; Lee, J.-S. Targeting Cancer Cells with Reactive Oxygen and Nitrogen Species Generated by Atmospheric-Pressure Air Plasma. *PLoS ONE* **2014**, *9*, e86173. [CrossRef] [PubMed]
2. Ahn, H.J.; Kim, K.I.; LEE, J.H.; Kim, J.H.; Yang, S.S.; Lee, J.-S. Cellular membrane collapse by atmospheric-pressure plasma jet. *Appl. Phys. Lett.* **2014**, *104*, 013701. [CrossRef]

3. Nguyen, N.H.; Park, H.J.; Yang, S.S.; Choi, K.S.; Lee, J.S. Anti-cancer efficacy of nonthermal plasma dissolved in a liquid, liquid plasma in heterogeneous cancer cells. *Sci. Rep.* **2016**, *6*, 29020. [CrossRef] [PubMed]

4. Kong, M.G.; Kroesen, G.; Morfill, G.; Nosenko, T.; Shimizu, T.; Dijk, J.; Zimmermann, J.L. Plasma medicine: An introductory review. *New J. Phys.* **2009**, *11*, 115012. [CrossRef]

5. Dobrynin, D.; Fridman, G.; Friedman, G.; Fridman, A. Physical and biological mechanisms of direct plasma interaction with living tissue. *New J. Phys.* **2009**, *11*, 115020. [CrossRef]

6. Dubuc, A.; Monsarrat, P.; Virard, F.; Merbahi, N.; Sarrette, J.-P.; Laurencin-Dalicieux, S. Use of cold-atmospheric plasma in oncology: A concise systemic review. *Ther. Adv. Med. Oncol.* **2018**, *10*, 1758835918786475. [CrossRef] [PubMed]

7. Liedtke, K.R.; Bekeschus, S.; Kaeding, A.; Hackbarth, C.; Kuehn, J.-P.; Heidecke, C.-D.; von Bernstorff, W.; von Woedtke, T.; Partecke, L.I. Non-thermal plasma-treated solution demonstrates antitumor activity against pancreatic cancer cells in vitro and in vivo. *Sci. Rep.* **2017**, *7*, 8319. [CrossRef] [PubMed]

8. Tanaka, H.; Mizuno, M.; Ishikawa, K.; Nakamura, K.; Kajiyama, H.; Kano, H.; Kikkawa, F.; Hori, M. Plasma-activated medium selectively kills glioblastoma brain tumor cells by down-regulating a survival signaling molecule, AKT kinase. *Plasma Med.* **2011**, *1*, 265–277. [CrossRef]

9. Kang, S.U.; Cho, J.-H.; Chang, J.W.; Shin, Y.S.; Kim, K.I.; Park, J.K.; Yang, S.S.; Lee, J.-S.; Moon, E.; Lee, K.; et al. Nonthermal plasma induces head and neck cancer cell death: The potential involvement of mitogen-activated protein kinase-dependent mitochondrial reactive oxygen species. *Cell Death Dis.* **2014**, *5*, e1056. [CrossRef] [PubMed]

10. Gweon, B.; Kim, M.; Kim, D.B.; Kim, D.; Kim, H.; Jung, H.; Shin, J.H.; Choe, W. Differential responses of human liver cancer and normal cells to atmospheric pressure plasma. *Appl. Phys. Lett.* **2011**, *99*, 063701. [CrossRef]

11. Hirst, A.M.; Simms, M.S.; Mann, V.M.; Maitland, N.J.; O'Connell, D.; Frame, F.M. Low temperature plasma treatment induces DNA damage leading to necrotic cell death in primary prostate epithelial cells. *Br. J. Cancer* **2015**, *112*, 1536–1545. [CrossRef] [PubMed]

12. Kalghatgi, S.; Kelly, C.M.; Cerchar, E.; Torabi, B.; Alekseev, O.; Fridman, A.; Friedman, G.; Azizkhan-Clifford, J. Effects of Non-Thermal Plasma on Mammalian Cells. *PLoS ONE* **2011**, *6*, e16270. [CrossRef] [PubMed]

13. Vandamme, M.; Robert, E.; Lerondel, S.; Sarron, V.; Ries, D.; Dozias, S.; Sobilo, J.; Gosset, D.; Kieda, C.; Legrain, B. ROS implication in a new antitumor strategy based on non-thermal plasma. *Int. J. Cancer.* **2012**, *130*, 2185. [CrossRef] [PubMed]

14. Iseki, S.; Nakamura, K.; Hayashi, M.; Tanaka, H.; Kondo, H.; Kajiyama, H.; Kano, H.; Kikkawa, F.; Hori, M. Selective killing of ovarian cancer cells through induction of apoptosis by nonequilibrium atmospheric pressure plasma. *Appl. Phys. Lett.* **2012**, *100*, 113702. [CrossRef]

15. Graves, D.B. The emerging role of reactive oxygen and nitrogen species in redox biology and some implications for plasma applications to medicine and biology. *J. Phys. D Appl. Phys.* **2012**, *45*, 263001. [CrossRef]

16. Gaunt, L.F.; Beggs, C.B.; Georghiou, G.E. Bactericidal Action of the Reactive Species Produced by Gas-Discharge Nonthermal Plasma at Atmospheric Pressure: A Review. *IEEE Trans. Plasma Sci.* **2006**, *34*, 1257–1269. [CrossRef]

17. Pancrazio, J.J.; Whelan, J.P.; Borkholder, D.A.; Ma, W.; Stenger, D.A. Development and Application of Cell-Based Biosensors. *Ann Biomed Eng.* **1999**, *27*, 697–711. [CrossRef] [PubMed]

18. Kim, K.; Choi, J.D.; Hong, Y.C.; Kim, G.; Noh, E.J.; Lee, J.-S.; Yang, S.S. Atmospheric-pressure plasma-jet from micronozzle array and its biological effects on living cells for cancer therapy. *Appl. Phys. Lett.* **2011**, *98*, 073701. [CrossRef]

19. Ahn, H.J.; Kim, K.I.; Kim, G.; Moon, E.; Yang, S.S.; Lee, J.-S. Atmospheric-Pressure Plasma Jet Induces Apoptosis Involving Mitochondria via Generation of Free Radicals. *PLoS ONE* **2011**, *6*, e28154. [CrossRef] [PubMed]

20. Adachi, T.; Tanaka, H.; Nonomura, S.; Hara, H.; Kondo, S.-I.; Hori, M. Plasma-activated medium induces A549 cell injury via a spiral apoptotic cascade involving the mitochondrial-nuclear network. *Free Radic. Biol. Med.* **2015**, *79*, 28–44. [CrossRef] [PubMed]

21. Kurake, N.; Tanaka, H.; Ishikawa, K.; Kondo, T.; Sekine, M.; Nakamura, K.; Kajiyama, H.; Kikkawa, F.; Mizuno, M.; Hori, M. Cell survival of glioblastoma grown in medium containing hydrogen peroxide and/or nitrite, or in plasma-activated medium. *Arch. Biochem. Biophys.* **2016**, *605*, 102–108. [CrossRef] [PubMed]

22. Girad, P.M.; Arbabian, A.; Fleury, M.; Bauville, G.; Puech, V.; Dutrix, M.; Sousa, J.S. Synergistic effect of H_2O_2 and NO_2^- in cell death induced by cold atmospheric He plasma. *Sci. Rep.* **2016**, *6*, 29098. [CrossRef] [PubMed]

23. Canal, C.; Fontelo, R.; Hamouda, I.; Guillem-Marti, J.; Cvelbar, U.; Ginebra, M.-P. Plasma-induced selectivity in bone cancer cell death. *Free Radic. Biol. Med.* **2017**, *110*, 72–80. [CrossRef] [PubMed]

24. Yan, D.; Nourmohammadi, N.; Bian, K.; Murad, F.; Sherman, J.H.; Keida, M. Stabilizing the cold plasma-stimulated medium by regulating medium's composition. *Sci. Rep.* **2016**, *6*, 26016. [CrossRef] [PubMed]

25. Yan, D.; Cui, H.; Zhu, W.; Nourmohammadi, N.; Milberg, J.; Zhang, L.G.; Sherman, J.H.; Keida, M. The specific vulnerabilities of cancer cells to the cold atmospheric plasma-stimulated solutions. *Sci. Rep.* **2017**, *7*, 4479. [CrossRef] [PubMed]

26. Lee, C.; Kim, T.; Park, H.; Yang, S.S. Stability improvement of nonthermal atmospheric-pressure plasma jet using electric field dispersion. *Microelectron. Eng.* **2015**, *145*, 153–159. [CrossRef]

27. Tichonovasab, M.; Kruglya, E.; Racysa, V.; Hipplerb, R.; Kaunelienea, V.; Stasiulaitienea, I.; Martuzevicius, D. Degradation of various textile dyes as wastewater pollutants under dielectric barrier discharge plasma treatment. *Chem. Eng.* **2013**, *229*, 9–19. [CrossRef]

28. Li, Y.; Kang, M.H.; Uhm, H.S.; Lee, G.J.; Choi, E.H.; Han, I. Effects of atmospheric-pressure non-thermal bio-compatible plasma and plasma activated nitric oxide water on cervical cancer cells. *Sci. Rep.* **2017**, *7*, 45781. [CrossRef] [PubMed]

Article

Activation of Murine Immune Cells upon Co-culture with Plasma-treated B16F10 Melanoma Cells

Katrin Rödder [1], Juliane Moritz [1], Vandana Miller [1,2], Klaus-Dieter Weltmann [1], Hans-Robert Metelmann [3], Rajesh Gandhirajan [1] and Sander Bekeschus [1,*]

[1] Centre for Innovation Competence (ZIK) *plasmatis*, Leibniz Institute for Plasma Science and Technology (INP Greifswald), Felix-Hausdorff-Str.2, 17489 Greifswald, Germany; katrin.roedder@inp-greifswald.de (K.R.); juliane.moritz@inp-greifswald.de (J.M.); vmiller@coe.drexel.edu (V.M.); weltmann@inp-greifswald.de (K.-D.W.); rajesh.gandhirajan@inp-greifswald.de (R.G.)

[2] Department of Microbiology and Immunology, Drexel University, 245 N. 15th Street, Philadelphia, PA 19102, USA

[3] Department of Oral and Maxillofacial Surgery/Plastic Surgery, Greifswald University Medical Center, Ferdinand-Sauerbruch-Str. DZ 7, 17475 Greifswald, Germany; metelmann@uni-greifswald.de

* Correspondence: sander.bekeschus@inp-greifswald.de; Tel.: +49-3834-554-3948

Received: 23 January 2019; Accepted: 12 February 2019; Published: 15 February 2019

Abstract: Recent advances in melanoma therapy increased median survival in patients. However, death rates are still high, motivating the need of novel avenues in melanoma treatment. Cold physical plasma expels a cocktail of reactive species that have been suggested for cancer treatment. High species concentrations can be used to exploit apoptotic redox signaling pathways in tumor cells. Moreover, an immune-stimulatory role of plasma treatment, as well as plasma-killed tumor cells, was recently proposed, but studies using primary immune cells are scarce. To this end, we investigated the role of plasma-treated murine B16F10 melanoma cells in modulating murine immune cells' activation and marker profile. Melanoma cells exposed to plasma showed reduced metabolic and migratory activity, and an increased release of danger signals (ATP, CXCL1). This led to an altered cytokine profile with interleukin-1β (IL-1β) and CCL4 being significantly increased in plasma-treated mono- and co-cultures with immune cells. In T cells, plasma-treated melanoma cells induced extracellular signal-regulated Kinase (ERK) phosphorylation and increased CD28 expression, suggesting their activation. In monocytes, CD115 expression was elevated as a marker for activation. In summary, here we provide proof of concept that plasma-killed tumor cells are recognized immunologically, and that plasma exerts stimulating effects on immune cells alone.

Keywords: kINPen; lymphocytes; macrophages; plasma medicine; reactive species

1. Introduction

Melanoma incidence rapidly increased over the last decades [1], and malignant melanoma is the most lethal form of skin cancer today [2]. Complicating its clinical treatment, the heterogeneity of malignant melanoma challenges the design of effective anti-tumor therapies [3]. Gold standard approaches in treatment and palliation include surgery, radiation therapy, chemotherapy, electrochemotherapy, adoptive cell transfer strategies, and immunotherapy [4–7]. The outstanding success of immunotherapies awarded with the Nobel Prize for Medicine and Physiology 2018 has provided compelling clinical evidence that the immune system plays a critical role in tumor defense [8]. Research interest in oncology is therefore not only focused on whether tumor cells are eliminated by a given substance or therapy, but also whether they can be recognized by immune cells [9]. Moreover, plasma treatment itself can also exert effects on immune cells [10–12].

The concept of immunogenic cancer cell death (ICD) proposes stimulation of an immune response against apoptotic tumor cells in a living host [13]. Apoptotic tumor cells, exposing plasma membrane-localized or intracellular molecules (damage associated molecular patterns, DAMPs), recruit myeloid cells [14]. Internalizing apoptotic cells, myeloid cells are capable to cross-present processed antigens on MHC class I molecules to tumor specific T cells [15]. This process results in a pro-inflammatory immune response [16], inducing tumor specific CD8$^+$ T cells potentially able to target melanoma cells [17]. Physical therapies in oncology have been shown to be potent ICD inducers, such as ionizing radiation, electrochemotherapy, and photodynamic therapy [18]. Another medical technology, cold physical plasma, was recently suggested for anti-melanoma therapy and danger signaling [19–21].

For medical purposes, cold physical plasma is a partially ionized gas operated at atmospheric pressure and body temperature [22]. It reacts with ambient air and generates RONS species leading to decreased melanoma cell activity and growth in vitro [23–25] and in vivo [26–28]. High species concentrations can be used to exploit apoptotic redox signaling pathways in tumor cells [29], and first cancer patients have benefited from plasma therapy [30]. Plasma treatment can act in concert with other drugs [31–33], and plasma-generated reactive species induce pro-immunogenic molecules on tumor cells, such as ecto-calreticulin (CRT) [34–36]. Although the release of this and other DAMP signals after plasma treatment in melanoma cells has been shown, evidence of immune cell activation in response to plasma-treated melanoma is scarce. We here performed co-cultures of plasma-treated murine melanoma cells with murine splenocytes consisting of myeloid as well as lymphoid cells. Plasma treatment decreased melanoma activity and increased release of danger signals. The cytokine profile as well as cellular activation markers and signaling pathways were investigated in immune cells to track a hypothesized perception of plasma-killed tumor cells by lymphocytes or phagocytes.

2. Materials and Methods

2.1. Cell Culture and Preparation of Splenocytes

Murine metastatic melanoma B16F10 cells were cultured in RPMI 1640 (PAN-Biotech, Germany) supplemented with 10% fetal bovine serum (FCS), 2% penicillin/streptomycin, and 1% glutamine (Sigma, Germany). A total of 2–4 $\times$ 10^4 cells were seeded in 1 ml cell culture medium in cell-culture treated 24-well plates one day prior to experimentation and allowed to adhere overnight. Spleens of eleven mice were prepared across all experiments using mechanical force and a 40 μm cell strainer (VWR, Germany) to obtain splenocytes suspension containing, for example, T cells and myeloid cells, including macrophages and dendritic cells (DCs). Cells were washed and red blood cells lysed with lysis buffer (BioLegend, USA) before washing and suspending splenocytes in fully supplemented culture medium. Sacrifice and tissue harvesting were performed in accordance with ethical standards.

2.2. Plasma Treatment and Co-Culture

Plasma treatment was performed using an atmospheric pressure non-thermal plasma released by the plasma jet kINPen (neoplas, Germany), which was shown to exert no mutagenic effects [37]. As feed gas, argon (99.9999% pure; Air Liquide, France) was used with a gas flow rate of 3 standard liters per minute (slm). Plasma was generated with a sinusoidal voltage of 2–6 kV$_{pp}$ with a frequency of ~1 MHz. The plasma jet was operated for 30 min prior to experiments to reduce effects of residual tube humidity on the plasma species composition [38]. The melanoma cells and splenocytes, respectively, were treated directly in 24-well plates. The plasma treatment of one well did not "spill over" into an adjacent (e.g., control) well as determined via measuring deposition of oxidants in both the treated and adjacent untreated wells in preliminary experiments (data not shown). The cultured well plate was placed 20 mm distant from the jet nozzle. The distance from the nozzle does play an important role as outlined previously [38]. We adjusted the distance in a way that the jet is as close as possible to the treated solution without liquid spilling into the jet (due to vortexing of the solution by the gas flux) that would extinguish the plasma ignition. The evaporation of the liquid induced by the gas flow of the kINPen was compensated for using a pre-determined amount of double-distilled water (to maintain iso-osmotic conditions) as described before [10]. For co-culture experiments, 1–2×10^6 splenocytes were added to each well one hour after plasma treatment of melanoma cells.

2.3. Quantification of Cytokines, Chemokines, and ATP

For quantification of 12 different murine cytokines or chemokines, bead-based multiplex analysis was performed according to the vendor's instructions (BioLegend). Supernatants were harvested 24 h (monocultures) or 4 h (trans-well co-culture) after plasma treatment, and stored at $-20\ ^\circ$C for one week until final analysis. In preliminary experiments, we could not observe a deteriorating effect on a single target due to storage, and any of such effect would apply as systemic error to all samples of the measurement, keeping the relative differences the same. Targets analyzed were interferon (IFN)-γ, interleukin (IL)-2, tumor growth factor (TGF)-β, tumor necrosis factor (TNF)-α, CCL2 (MCP-1), CXCL9 (MIG), IL-10, IL-6, CCL4 (MIP-1β), IL-4, IL-12p70, and IL-1β. Bead fluorescence was quantified on a CytoFlex S flow cytometer (Beckman-Coulter, USA), and target concentration was quantified against a known standard supplied by the manufacturer at a given concentration. The mean fluorescence intensity of each dilution series of the standard aids in generating standard curves that allow the absolute quantification of each target investigated in the cell culture supernatant. For quantification of CXCL1 (KC) in supernatants 6h after plasma treatment, enzyme-linked immunosorbent assay (ELISA) was used (BioLegend) according to the manufacturer's protocol. ATP was measured in supernatants 6h after treatment using the ATP determination kit (Thermo Fisher Scientific) according to the manufacturer's protocol.

2.4. Quantification of H_2O_2

Quantification of total H_2O_2 concentrations was performed using Amplex UltraRed reagent (Thermo Fisher Scientific). The assay was performed in 96-well microplates (NUNC, Denmark) according to the manufacturers' protocol. Fluorescence was read using a microplate reader (Tecan, Switzerland) with appropriate filter settings (λ_{ex} 535 nm λ_{em} 590 nm). H_2O_2 levels were quantified against a known standard (dilution series of chemically produced H_2O_2 by Sigma, Germany).

2.5. Metabolic Activity

Metabolic activity was assessed by measuring the reduction of non-fluorescent resazurin (Thermo Fisher Scientific; final concentration of 100µM) to fluorescent resorufin after 4 h of incubation under cell culture conditions. In experiments using splenocytes, cells were incubated either with or without PMA (phorbol 12-myristate-13-acetate; Sigma, Germany) at a final concentration of 100 ng/ml. Fluorescence was read using a microplate reader (Tecan, Switzerland) with appropriate filter settings (λ_{ex} 535 nm λ_{em} 590 nm).

2.6. Scratch Assay

The scratch assay is a method to measure cell migration in vitro. The methodical basis contains a "scratch" in a cellular grown monolayer, monitoring the cellular regrowth until closure of the scratch. Essentially, the cellular migration is imaged and quantified at different time points after the scratch [39]. B16F10 melanoma cells were seeded at 7.5×10^4 cells per well in 24-well plates, yielding a confluent monolayer the next day. Melanoma cells were exposed to plasma or were left untreated, prior to scratching the monolayer with a 200 µl pipette tip vertically across the center of the well. The resulting gap distance was imaged (Observer Z.1; Zeiss, Germany) and quantified (AxioVision 4.91 software; Zeiss) at 0 h and 24 h. In between, cells were cultured under standard conditions.

2.7. Cell Surface Marker Expression

Splenocytes were incubated with Fc-block (BioLegend) to block non-specific binding of antibodies. In some experiments, cells were incubated for 20 min with Fluo-3 (ThermoFisher Scientific). To assess subpopulations and activation marker, the following anti-murine monoclonal and fluorescently tagged antibodies were used: CD3 Alexa Fluor 700, CD11b PE-Dazzle, CD115 Brilliant Violet 421, CD43 APC, CD44 PE-Dazzle 594, CD4 PE-Cy 7, CD8a PerCP/Cy 5.5, CD28 APC, CD45R APC-Cy7, I-A/I-E Alexa Fluor 488, Ly-6C Alexa Fluor 700, CD68 PerCP/Cy 5.5, Ly-6G Brilliant Violet 510, FcεRIα PE, CD86 PE-Cy7 (all BioLegend). For kinase activation assays, co-cultured splenocytes were fixed in methanol for 1 h at $-20\,^\circ$C, washed in PBS, and incubated for 20 min with monoclonal antibodies directed against murine phospho-Erk1/2 PerCP/eFluor 710 (eBioscience, Germany) and lineage markers. Multicolor flow cytometry data were analyzed with Kaluza 2.1.1 software (Beckman-Coulter).

2.8. Statistical Analysis

Statistical analysis was performed using prism 8.01 (GraphPad Software, USA). Mean and standard errors are given. Statistical significance was reported with asterisks (n.s. $p > 0.05$; * $p < 0.05$; ** $p < 0.01$; *** $p < 0.001$).

3. Results

3.1. Plasma-Deposited Oxidants Decreased Melanoma Cell Viability and Motility

Cold physical plasma introduces oxidants into liquids surrounding cells, with hydrogen peroxide (H_2O_2) being among the long-lasting products of redox chemistry [40]. In pure PBS (Figure 1a) and fully supplemented cell culture medium (Figure 1b), H_2O_2 accumulated in a dose-dependent manner, validating the introduction of plasma-derived species in culture systems. The concentration of H_2O_2 differed only modestly between complex cell culture medium compared to buffered saline (PBS), suggesting that FCS in cell culture medium has a negligible effect on H_2O_2 pre-cursor species from the kINPen. Furthermore, the effect of FCS as a growth factor for B16F10 was measured via metabolic activity using different FCS concentrations. As expected, the melanoma cell metabolic activity increased with increasing FCS concentration (Figure 1c). Following this, the effect of plasma treatment on melanoma growth with and without FCS was analyzed and not found to be substantially different at 4 h (Figure 1d). This suggests a growth-supporting but not a dramatic antioxidative effect caused by the proteins of FCS with plasma treatment. To see whether FCS presence or absence during plasma treatment affects the migratory capability in melanoma cells, scratch assays were performed (Figure 1e). Results showed that the presence FCS had only a minor impact on cellular migration, in contrast to plasma treatment, which reduced cell regrowth with increasing treatment times (Figure 1f). Importantly, FCS-containing medium was added after treatment to both conditions to focus on short-term effects following plasma treatment. This suggests that within our experimental system, FCS (composed of various proteins) in solution only played a minor role in mediating or protecting from plasma effects on tumor cells. Altogether, plasma treatment slowed melanoma cells' metabolic activity and motility, regardless of FCS presence or absence.

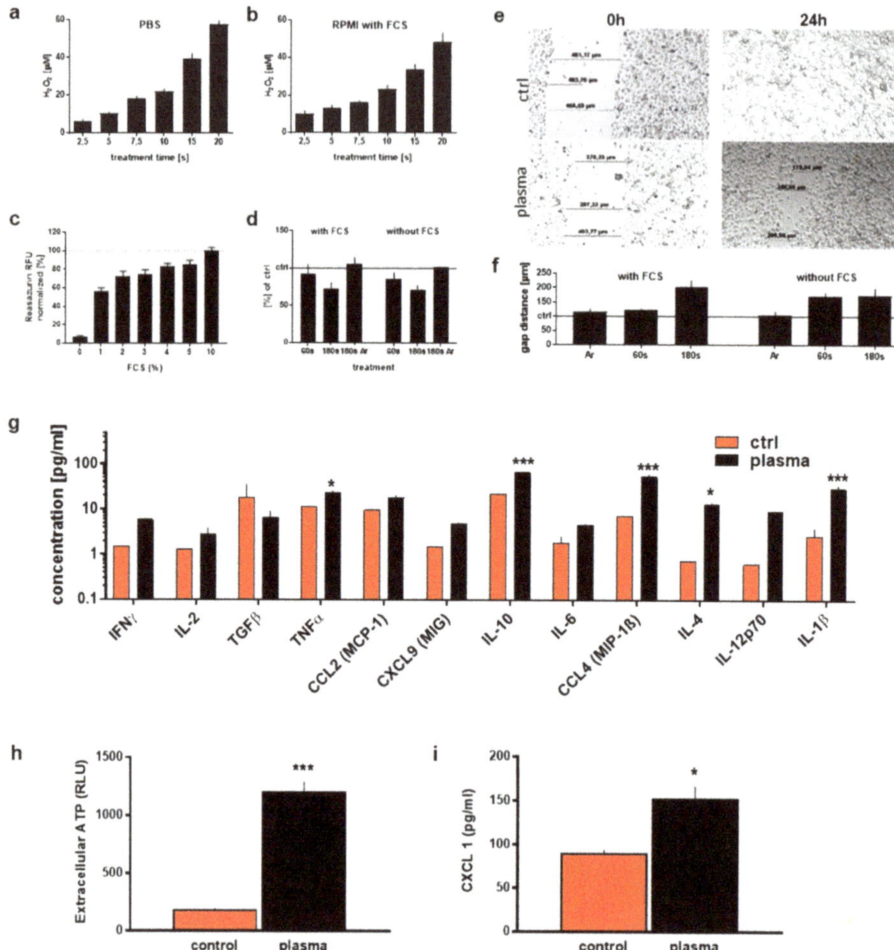

Figure 1. Effects of plasma treatment in liquids and treatment of B16F10 melanoma cells with physical plasma. (**a**) Plasma deposited hydrogen peroxide in PBS and (**b**) cell culture medium at nearly similar levels; (**c**) growth behavior of melanoma cells under different concentration of FCS; (**d**) reduction of metabolic activity in melanoma cells after plasma treatment in cell culture medium with or without FCS, normalization was done with respect to a control in each condition; (**e**) representative images of scratches in control or 180 s plasma-treated melanoma cells after 0 h or 24 h, respectively, and (**f**) their normalized regrowth quantified; (**g**) cytokine/chemokine profile of plasma-treated melanoma cells; (**h**) release of the danger molecules ATP and (**i**) CXCL1 in plasma-treated melanoma cells. (**a–d**,**f**) Data are mean +S.E. from several replicates; statistical analysis was performed using t-test; PBS: phosphate-buffered saline; FCS: fetal calf serum.

3.2. Plasma Treatment Modulated Inflammatory Milieu of Melanoma and Immune Cells

In response to 120 s plasma treatment, melanoma cell supernatants (analyzed after 24 h) showed significantly increased levels of TNFα, IL-10, CCL4, and IL-1β compared to untreated control cells (Figure 1g). An increase of damage-associated molecules, such as ATP (Figure 1h) and CXCL1 (Figure 1i), was also observed 6 h after a 120 s plasma-treatment in B16F10 cells. Before investigating the immunological consequences of murine splenocytes co-cultured with plasma-treated melanoma cells, the response of splenocytes alone treated with plasma was investigated. For this, splenocytes were harvested from mice (Figure 2a) and left unstimulated or were pulsed with PMA. Lymphocyte activation is typically triggered by the interaction of the cell surface receptor to its specific ligand, resulting in an activation cascade. In vitro, lymphocyte activation can be induced by chemicals, such as PMA, leading to an increased cellular proliferation. Following this, splenocytes were exposed to plasma, and metabolic activity was assessed 24 h later. Reduced metabolic activity was observed in both naïve and PMA-stimulated splenocytes, but in the latter to a lesser extent (Figure 2b). Studying the splenocytes' supernatants, increased levels of cytokines were observed after plasma treatment, including IL-10, CCL4, IL-4, IL-12, and IL-1β. Altogether, plasma treatment not only affected cell viability but also the inflammatory profile of tumor and immune cells (Figure 2c).

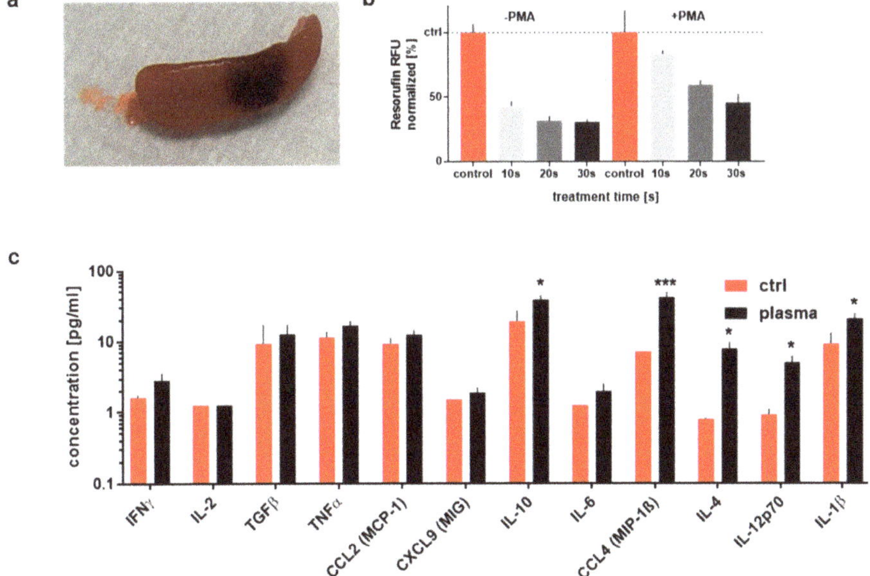

Figure 2. Treatment of murine immune cells with physical plasma. (**a**) Representative murine spleen after harvesting and prior to homogenization; (**b**) metabolic activity of naïve (-PMA) and activated (+PMA) splenocytes after plasma treatment; (**c**) cytokine/chemokine profile 24 h after 120 s plasma treatment in naïve splenocytes. Data are mean +S.E. of several spleens, statistical analysis was performed using t test. PMA: phorbol 12-myristate 13-acetate.

3.3. Activation of Immune Cells with Plasma Treatment

Cellular activation can be assessed using several methods such as calcium signaling. Fluo-3 is a labelled calcium indicator, and its fluorescence increases 100-fold upon calcium binding. Calcium influxes were observed in early lymphocyte activation. Total cellular fluorescence was measured by flow cytometry in macrophages (Figure 3a) and T cells (Figure 3b). After addition of Fluo-3 labeled splenocytes to plasma-treated melanoma cells, there was no significant change in calcium influx in macrophages (Figure 3c) and T cells (Figure 3d). Of note, a significant calcium influx was demonstrated for splenocytic T cells directly exposed to plasma and over time (Figure 3c) but not in macrophages (Figure 3d). Calcium release and protein kinase activity are closely linked. Being a member of the Ras-Raf-Erk signal transduction cascade, the Erk1/2 (extracellular signal-regulated Kinase-1/2) kinases are activated via phosphorylation. Upon activation, the Erk kinases function in cellular proliferation, differentiation, and survival. A significant increase (Figure 3e) of phosphorylated Erk1/2 was demonstrated for plasma-treated T cells after 1 h and 4 h (Figure 3f). For splenocytes co-cultured with plasma-treated B16F10 melanoma cells, a small but significant increase was observable 1 h after culture onset (Figure 3f). This suggests that plasma treatment and plasma-treated melanoma cells may be able to stimulate immune cells.

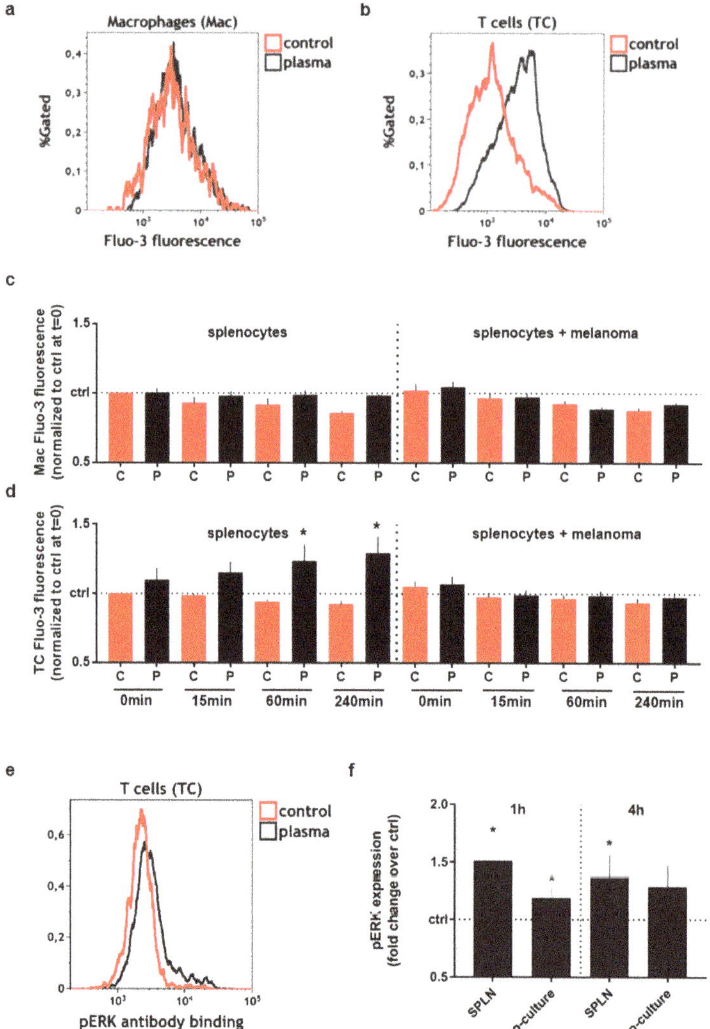

Figure 3. Activation of selected splenocyte subpopulations after co-cultivation with 120 s plasma-cells. (**a**) Representative overlay of calcium release assessed via flow cytometry in macrophages and (**b**) T cells, respectively; (**c**,**d**) quantification of calcium release over time in macrophages and T cells, respectively after plasma treatment (**e**) representative overlay of phosphorylated extracellular signal-regulated Kinase-1/2 (Erk1/2) kinase in T cells 1 h as measured via flow cytometry; (**f**) quantification of Erk1/2 phosphorylation in splenocytes with or without co-culture with 120 s plasma-treated melanoma cells after 1 h or 4 h. Data are mean +S.E. of several spleens; c=control, p=plasma, SPLN=splenocytes.

3.4. Co-Culture with Plasma-Treated Melanoma Cells Altered the Surface Marker and Cytokine Profile of Murine Immune Cells

Among splenocytes, CD4$^+$ T helper and CD8$^+$ cytotoxic T cells were identified from CD3$^+$ splenic lymphocytes (Figure 4a). With both lymphocyte subpopulations, the expression of the activation-indicating surface marker CD28 was assessed after co-culture with either untreated melanoma cells or plasma-treated melanoma cells (Figure 4b-c). CD28 expression on the cell surface increased in tendency for CD4$^+$ and significantly for CD8$^+$ T cells (Figure 4d). The CD28 cell surface marker provides essential co-stimulation during T cell activation. The expression of other activating T cell markers (CD44, CD152) did not change (data not shown). Only viable cells were included in the analysis, and we also investigated changes of the CD4/CD8 ratio. This parameter is used to describe the immune status, and we found an increase of CD4 over CD8 cells upon co-culture with plasma-treated over non-treated melanoma cells (Figure 4e). This suggests splenic CD4$^+$ T helper cells survive better in the microenvironment generated by plasma-treated melanoma cells. To assess any changes of myeloid immune cells under these conditions, splenocytes were gated for monocytes and macrophages (Figure 5a). Several surface markers for M1 macrophages (CD115), M2 macrophages (FcεR1), and co-stimulation during antigen presentation (CD86) were investigated (Figure 5b).

Monocytes had upregulated surface marker expression after co-culture with plasma-treated melanoma cells, with a significant increase of CD115, arguing for an onset of monocyte-to-macrophage maturation. Macrophages did not show any significant changes, with a non-significant decrease of CD86. Finally, cytokine analysis of splenocyte-melanoma trans-well co-cultures revealed increased levels of IL-10 and CCL4, with a trend of increased release with IL-1β, IL-12p70, TNFα, and TGFβ (Figure 5c). This argued for a modulation of the immune cells inflammatory response in when exposed to plasma-treated tumor cells.

4. Discussion

Our aim was to identify activation signatures in murine immune cells derived from spleen that had been co-cultured with plasma-treated melanoma cells. Although many responses observed in immune cells were rather subtle, their sum point to a principle recognition of plasma-treated compared to control melanoma cells.

Direct plasma treatment induced pro-inflammatory cytokine release of IL-1β or TNF-α in B16F10 melanoma cells. IL-1β is processed and released under caspase-8 regulated apoptotic cell death [41]. Apoptotic cells release uric acid, activating systemic inflammation and resulting in release of IL-1β [42]. Interleukin-1β is a potent activator molecule in T and B cell responses and supports survival of naive and memory T cells [43–45], potentially forming anti-tumor immunity. Tumor necrosis factor alpha (TNFα) leads to pyrogenic activation, inducing apoptotic cell death via the extrinsic pathway [46]. Plasma treatment of melanoma cells in mono and co-culture also increased release of IL-10 and CCL, both positively regulating T cell function [47]. Hence, plasma-derived oxidative cell stress and/or death of melanoma cells profoundly affected inflammatory conditioning of cells of the immune system. Presence of FCS during plasma treatment was of minor importance corroborating previous results that excess protein has negligible effects on plasma-derived long-lived oxidants [36]. Plasma-induced tumor cell death supposedly was of an immunogenic nature, as release of ATP and CXCL1 suggests [48], which is in line with earlier reports [34,35,49,50].

In this study, immune cells derived from spleens were exposed to plasma. Similar to studies with human immune cells [51–53], splenocytes' viability was affected by direct exposure to plasma, while their mitogenic activation (with PMA) reduced plasma-mediated toxic effects. As the conditions (cell number, treatment time, volume of media, etc.) were the same in both non-activated and activated regimens but with different outcomes, our results point to differences in intracellular signal translation upon expose to reactive species, as these responses ultimately determines the amplitude of toxicity. At lower concentrations, reactive species can also serve in cell signaling [54].

Several studies manifested the role of ROS, in particular H_2O_2, as a stimulant and second messenger in lymphocytes [55]. An increase in intracellular calcium was found in plasma-treated T cells in our study, which points to their activation [56,57]. Elevated intracellular calcium levels need to be maintained for more than 30 min for T cell activation [57,58], which was the case with our results. Moreover, we determined a significant increase in Erk1/2 activation in T cells, which is associated with activation [59,60]. In general, we do no propose a specific but rather non-specific stimulus of lymphocytes with plasma as this is highly unlikely without proper T cell receptor stimulation as previously described [22] and suggested experimentally [61,62]. Co-culture of splenocytes with plasma-treated melanoma cells provoked calcium influx neither in T cells nor in macrophages. For T cells, this suggests a lack of specific antigen being presented either on tumor (MHC I) or antigen presenting cells (MHC II). For macrophages, calcium signaling is often related to aberrant stress of the endoplasmic reticulum and cell death [63] or sensing of microorganisms [64].

Both $CD4^+$ T helper and $CD8^+$ cytotoxic T cells contribute to antitumor immunity in tumor patients [65–67]. $CD4^+$ and $CD8^+$ T cells co-cultured with plasma-treated compared to untreated melanoma cells showed an increased expression of CD28. When binding to its ligands (B7 receptors, e.g., CD80/86) triggering IL-2 release [68], the CD28 costimulatory receptor is required for efficient and optimal immune response in T lymphocytes [69]. CD28-signaling through Bcl-2 and Bcl-xL promotes T cell survival [70–72]. Importantly, basal CD28 expression in healthy volunteers, as well as in patients, correlates with a T cell phenotype more sensitive to stimulation, and low CD28 expression is a marker of disease [73]. Moreover, plasma treatment of splenocytes significantly increased levels of IL-4 and IL-12. The former induces T_H2 and the latter T_H1 responses, having opposite roles in health and disease [74]. Both cytokines are derived from myeloid cells such as basophils and monocytes/macrophages [75]. Although calcium increase was not observed in immune cells upon co-cultured with plasma-treated melanoma cells, an upregulation of surface marker was observed in monocytes in that condition. CD86 expression is induced upon stimulation in monocytes, leading to an inflammatory milieu [76] and proper T cell co-stimulation [77]. Increase of IgE receptor (FcεR) dependent activation of monocytes is associated with tissue inflammation and cytotoxicity allergic diseases and asthma [78,79]. However, significantly regulated was only colony-stimulating factor 1 receptor (CSF-1R or CD115). CD115 contributes to macrophage maturation [80] and is crucial for the development of certain myeloid subsets as CD115 blockage shortens the life span of monocytes [81].

Several studies by others and us have observed plasma-induced tumor cell death before [82–98]. Kumar and colleagues observed in several pancreatic cancer cell lines an increase in apoptosis concomitant with downregulation of MAPK7, BCL2, and CHK1 [99]. In line with our results, they have observed an increase in H_2O_2 with plasma treatment using the kINPen, and moreover found apoptosis-inducing effects of plasma treatment in lung cancer cells based on increase in plasma-mediated ROS formation and intracellular oxidation signaled via ATM and p53 machinery [100], which is supported by our recent data in HaCaT keratinocytes [101].

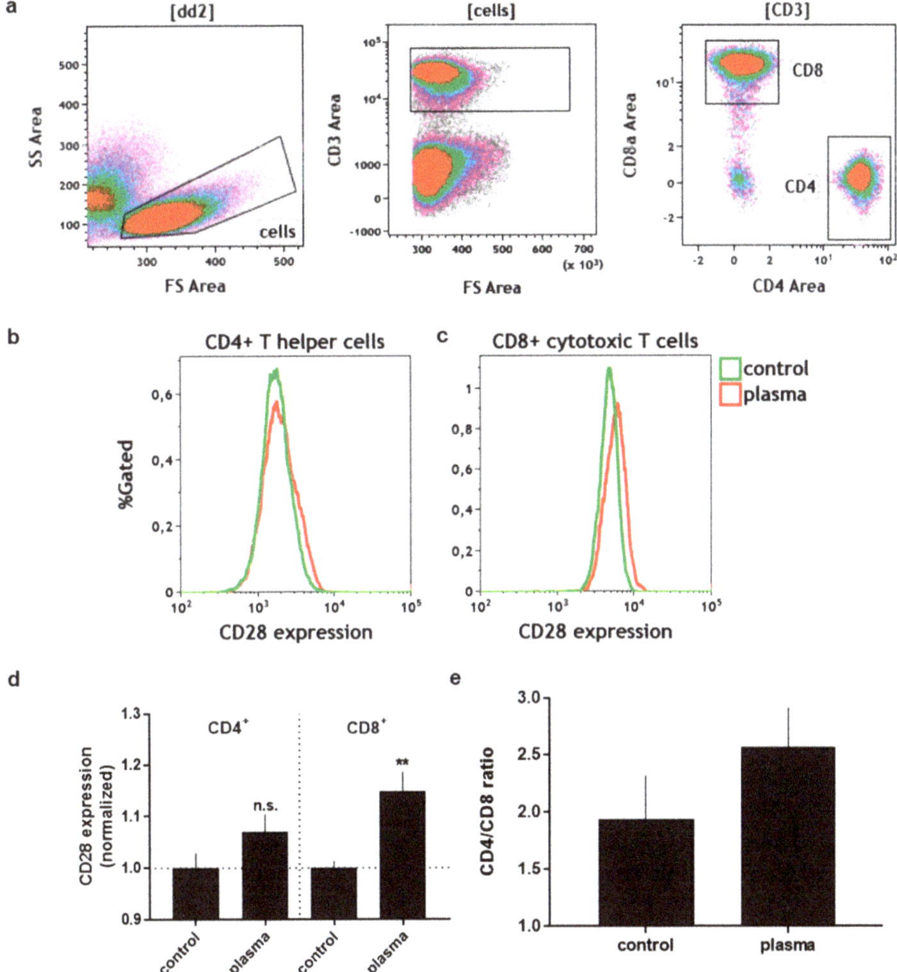

Figure 4. CD28 expression in T lymphocyte splenocyte subpopulations co-cultured with control or plasma-treated melanoma cells. (**a**) Representative flow cytometric dot plots used to gate for two T lymphocyte subpopulations, namely CD4+ T helper and CD8+ cytotoxic T cells; (**b**,**c**) representative overlay histograms for CD28 marker in expression in CD4+ and CD8+ cells, respectively; (**d**) fold change quantification of CD28 expression in T cells co-cultured for 24 h with either control or plasma-treated melanoma cells; (**e**) CD4/CD8 of T cells co-cultured for 24 h with either control or plasma-treated melanoma cells. Data are mean + S.E. of several spleens, statistical analysis was performed using t test.

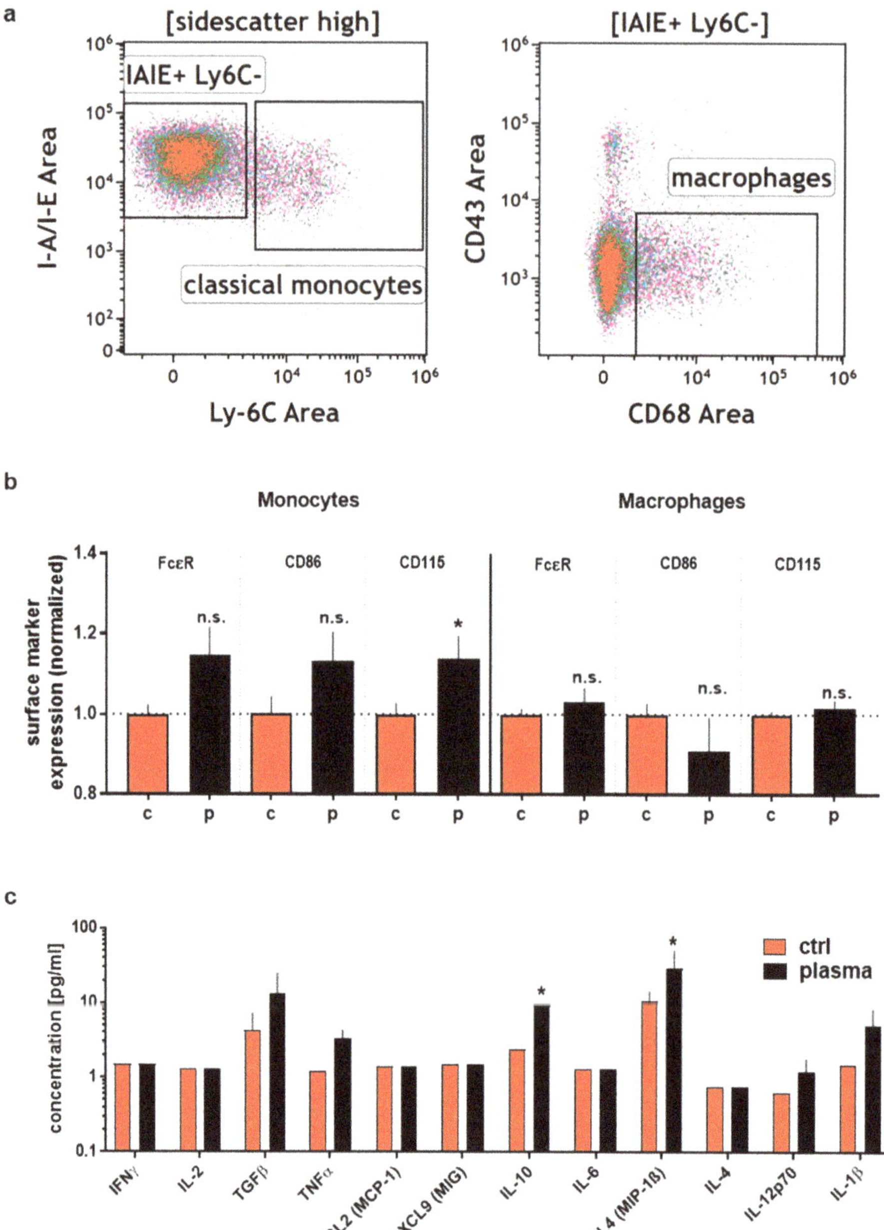

Figure 5. Activation of monocyte and macrophage splenocyte subpopulations after co-culture with melanoma cells. (**a**) Representative gating strategy to identify monocytes and macrophages within splenocytes; (**b**) of change expression of activation marker expression in monocytes/macrophages 24 h after co-culture with either control (C) or plasma-treated (P) melanoma cells; (**c**) cytokine profile supernatants retrieved from transwell co-cultures of control or plasma-treated melanoma cells with splenocytes. Data are presented as mean + S.E. of several spleens; statistical analysis was performed using t test.

5. Conclusions

Our findings indicate a tumor-static action in terms of metabolic activity and cell motility of plasma on melanoma cells and—in our hands—a negligible protective effect of protein present during the treatment. We further observed a role of plasma-mediated activation of splenic immune cells, as well as an effect of plasma-treated melanoma cells on immune cells. Specifically, plasma treatment modulated inflammatory parameters (such as cytokines and cell surface activation markers), which argue for pro-immunogenic role of plasma treatment. Additional studies with further differentiated assays and immune phenotyping are needed to understand how plasma technology can be used in onco-immunology to decrease tumor burden.

Author Contributions: Conceptualization, S.B. and V.M.; methodology, S.B., K.R., J.M. and R.G.; formal analysis, S.B., K.R., J.M. and R.G.; investigation, S.B., K.R. and J.M.; resources, S.B.; writing—original draft preparation, S.B.; writing—review and editing, S.B., V.M., K.-D.W. and H.-R.M.; supervision, S.B. and H.-R.M.; project administration, S.B.; funding acquisition, S.B. and K.-D.W.

Funding: This research was funded by the German Federal Ministry of Education and Research (BMBF), grant number 03Z22DN11.

Conflicts of Interest: The authors declare no conflict of interest.

References

1. Linos, E.; Swetter, S.M.; Cockburn, M.G.; Colditz, G.A.; Clarke, C.A. Increasing burden of melanoma in the United States. *J. Investig. Dermatol.* **2009**, *129*, 1666–1674. [CrossRef] [PubMed]
2. Erdei, E.; Torres, S.M. A new understanding in the epidemiology of melanoma. *Expert Rev. Anticancer Ther.* **2010**, *10*, 1811–1823. [CrossRef] [PubMed]
3. Bittner, M.; Meltzer, P.; Chen, Y.; Jiang, Y.; Seftor, E.; Hendrix, M.; Radmacher, M.; Simon, R.; Yakhini, Z.; Ben-Dor, A.; et al. Molecular classification of cutaneous malignant melanoma by gene expression profiling. *Nature* **2000**, *406*, 536–540. [CrossRef]
4. Bajetta, E.; Del Vecchio, M.; Bernard-Marty, C.; Vitali, M.; Buzzoni, R.; Rixe, O.; Nova, P.; Aglione, S.; Taillibert, S.; Khayat, D. *Metastatic Melanoma: Chemotherapy*; Elsevier: Amsterdam, The Netherlands, 2002; pp. 427–445.
5. Jandus, C.; Speiser, D.; Romero, P. Recent advances and hurdles in melanoma immunotherapy. *Pigment Cell Melanoma Res.* **2009**, *22*, 711–723. [CrossRef] [PubMed]
6. Queirolo, P.; Marincola, F.; Spagnolo, F. Electrochemotherapy for the management of melanoma skin metastasis: A review of the literature and possible combinations with immunotherapy. *Arch. Dermatol. Res.* **2014**, *306*, 521–526. [CrossRef] [PubMed]
7. Rosenberg, S.A.; Dudley, M.E. Adoptive cell therapy for the treatment of patients with metastatic melanoma. *Curr. Opin. Immunol.* **2009**, *21*, 233–240. [CrossRef] [PubMed]
8. Banchereau, J.; Palucka, K. Immunotherapy: Cancer vaccines on the move. *Nat. Rev. Clin. Oncol.* **2018**, *15*, 9–10. [CrossRef] [PubMed]
9. Garg, A.D.; Romano, E.; Rufo, N.; Agostinis, P. Immunogenic versus tolerogenic phagocytosis during anticancer therapy: Mechanisms and clinical translation. *Cell Death Differ.* **2016**, *23*, 938–951. [CrossRef] [PubMed]
10. Bekeschus, S.; Moritz, J.; Schmidt, A.; Wende, K. Redox regulation of leukocyte-derived microparticle release and protein content in response to cold physical plasma-derived oxidants. *Clin. Plasma Med.* **2017**, *7–8*, 24–35. [CrossRef]
11. Bekeschus, S.; Clemen, R.; Metelmann, H.-R. Potentiating anti-tumor immunity with physical plasma. *Clin. Plasma Med.* **2018**, *12*, 17–22. [CrossRef]
12. Bekeschus, S.; Seebauer, C.; Wende, K.; Schmidt, A. Physical plasma and leukocytes—Immune or reactive? *Biol. Chem.* **2019**, *400*, 63–75. [CrossRef] [PubMed]
13. Green, D.R.; Ferguson, T.; Zitvogel, L.; Kroemer, G. Immunogenic and tolerogenic cell death. *Nat. Rev. Immunol.* **2009**, *9*, 353–363. [CrossRef] [PubMed]

14. Obeid, M.; Tesniere, A.; Ghiringhelli, F.; Fimia, G.M.; Apetoh, L.; Perfettini, J.L.; Castedo, M.; Mignot, G.; Panaretakis, T.; Casares, N.; et al. Calreticulin exposure dictates the immunogenicity of cancer cell death. *Nat. Med.* **2007**, *13*, 54–61. [CrossRef]

15. Ghiringhelli, F.; Apetoh, L.; Tesniere, A.; Aymeric, L.; Ma, Y.; Ortiz, C.; Vermaelen, K.; Panaretakis, T.; Mignot, G.; Ullrich, E. Activation of the nlrp3 inflammasome in dendritic cells induces il-1β–dependent adaptive immunity against tumors. *Nat. Med.* **2009**, *15*, 1170. [CrossRef] [PubMed]

16. Kroemer, G.; Galluzzi, L.; Kepp, O.; Zitvogel, L. Immunogenic cell death in cancer therapy. *Annu. Rev. Immunol.* **2013**, *31*, 51–72. [CrossRef]

17. Nowak, A.K.; Lake, R.A.; Marzo, A.L.; Scott, B.; Heath, W.R.; Collins, E.J.; Frelinger, J.A.; Robinson, B.W. Induction of tumor cell apoptosis in vivo increases tumor antigen cross-presentation, cross-priming rather than cross-tolerizing host tumor-specific cd8 t cells. *J. Immunol.* **2003**, *170*, 4905–4913. [CrossRef] [PubMed]

18. Adkins, I.; Fucikova, J.; Garg, A.D.; Agostinis, P.; Spisek, R. Physical modalities inducing immunogenic tumor cell death for cancer immunotherapy. *Oncoimmunology* **2014**, *3*, e968434. [CrossRef] [PubMed]

19. Pasqual-Melo, G.; Gandhirajan, R.K.; Stoffels, I.; Bekeschus, S. Targeting malignant melanoma with physical plasmas. *Clin. Plasma Med.* **2018**, *10*, 1–8. [CrossRef]

20. Miller, V.; Lin, A.; Fridman, A. Why target immune cells for plasma treatment of cancer. *Plasma Chem. Plasma Process.* **2015**, *36*, 259–268. [CrossRef]

21. Barekzi, N.; Laroussi, M. Effects of low temperature plasmas on cancer cells. *Plasma Process. Polym.* **2013**, *10*, 1039–1050. [CrossRef]

22. Bekeschus, S.; Favia, P.; Robert, E.; von Woedtke, T. White paper on plasma for medicine and hygiene: Future in plasma health sciences. *Plasma Process. Polym.* **2019**, *16*, 1800033. [CrossRef]

23. Schmidt, A.; Bekeschus, S.; von Woedtke, T.; Hasse, S. Cell migration and adhesion of a human melanoma cell line is decreased by cold plasma treatment. *Clin. Plasma Med.* **2015**, *3*, 24–31. [CrossRef]

24. Arndt, S.; Wacker, E.; Li, Y.F.; Shimizu, T.; Thomas, H.M.; Morfill, G.E.; Karrer, S.; Zimmermann, J.L.; Bosserhoff, A.K. Cold atmospheric plasma, a new strategy to induce senescence in melanoma cells. *Exp. Dermatol.* **2013**, *22*, 284–289. [CrossRef] [PubMed]

25. Choi, B.B.R.; Choi, J.H.; Hong, J.W.; Song, K.W.; Lee, H.J.; Kim, U.K.; Kim, G.C. Selective killing of melanoma cells with non-thermal atmospheric pressure plasma and p-fak antibody conjugated gold nanoparticles. *Int. J. Med. Sci.* **2017**, *14*, 1101–1109. [CrossRef] [PubMed]

26. Binenbaum, Y.; Ben-David, G.; Gil, Z.; Slutsker, Y.Z.; Ryzhkov, M.A.; Felsteiner, J.; Krasik, Y.E.; Cohen, J.T. Cold atmospheric plasma, created at the tip of an elongated flexible capillary using low electric current, can slow the progression of melanoma. *PLoS ONE* **2017**, *12*, e0169457. [CrossRef] [PubMed]

27. Daeschlein, G.; Scholz, S.; Lutze, S.; Arnold, A.; von Podewils, S.; Kiefer, T.; Tueting, T.; Hardt, O.; Haase, H.; Grisk, O.; et al. Comparison between cold plasma, electrochemotherapy and combined therapy in a melanoma mouse model. *Exp. Dermatol.* **2013**, *22*, 582–586. [CrossRef] [PubMed]

28. Iida, M.; Omata, Y.; Nakano, C.; Yajima, I.; Tsuzuki, T.; Ishikawa, K.; Hori, M.; Kato, M. Decreased expression levels of cell cycle regulators and matrix metalloproteinases in melanoma from ret-transgenic mice by single irradiation of non-equilibrium atmospheric pressure plasmas. *Int. J. Clin. Exp. Pathol.* **2015**, *8*, 9326–9331.

29. Lorenzen, I.; Mullen, L.; Bekeschus, S.; Hanschmann, E.M. Redox regulation of inflammatory processes is enzymatically controlled. *Oxid. Med. Cell. Longev.* **2017**, *2017*, 8459402. [CrossRef]

30. Metelmann, H.R.; Seebauer, C.; Rutkowski, R.; Schuster, M.; Bekeschus, S.; Metelmann, P. Treating cancer with cold physical plasma: On the way to evidence-based medicine. *Contrib. Plasma Phys.* **2018**, *58*, 415–419. [CrossRef]

31. Gandhirajan, R.K.; Rodder, K.; Bodnar, Y.; Pasqual-Melo, G.; Emmert, S.; Griguer, C.E.; Weltmann, K.D.; Bekeschus, S. Cytochrome c oxidase inhibition and cold plasma-derived oxidants synergize in melanoma cell death induction. *Sci. Rep.* **2018**, *8*, 12734. [CrossRef]

32. Sagwal, S.K.; Pasqual-Melo, G.; Bodnar, Y.; Gandhirajan, R.K.; Bekeschus, S. Combination of chemotherapy and physical plasma elicits melanoma cell death via upregulation of slc22a16. *Cell Death Dis.* **2018**, *9*, 1179. [CrossRef] [PubMed]

33. Masur, K.; von Behr, M.; Bekeschus, S.; Weltmann, K.D.; Hackbarth, C.; Heidecke, C.D.; von Bernstorff, W.; von Woedtke, T.; Partecke, L.I. Synergistic inhibition of tumor cell proliferation by cold plasma and gemcitabine. *Plasma Process. Polym.* **2015**, *12*, 1377–1382. [CrossRef]

34. Bekeschus, S.; Rodder, K.; Fregin, B.; Otto, O.; Lippert, M.; Weltmann, K.D.; Wende, K.; Schmidt, A.; Gandhirajan, R.K. Toxicity and immunogenicity in murine melanoma following exposure to physical plasma-derived oxidants. *Oxid. Med. Cell. Longev.* **2017**, *2017*, 4396467. [CrossRef] [PubMed]

35. Lin, A.; Truong, B.; Patel, S.; Kaushik, N.; Choi, E.H.; Fridman, G.; Fridman, A.; Miller, V. Nanosecond-pulsed dbd plasma-generated reactive oxygen species trigger immunogenic cell death in a549 lung carcinoma cells through intracellular oxidative stress. *Int. J. Mol. Sci.* **2017**, *18*, 966. [CrossRef] [PubMed]

36. Bekeschus, S.; Schmidt, A.; Niessner, F.; Gerling, T.; Weltmann, K.D.; Wende, K. Basic research in plasma medicine—A throughput approach from liquids to cells. *J. Vis. Exp.* **2017**, e56331. [CrossRef] [PubMed]

37. Bekeschus, S.; Schmidt, A.; Kramer, A.; Metelmann, H.R.; Adler, F.; von Woedtke, T.; Niessner, F.; Weltmann, K.D.; Wende, K. High throughput image cytometry micronucleus assay to investigate the presence or absence of mutagenic effects of cold physical plasma. *Environ. Mol. Mutagen.* **2018**, *59*, 268–277. [CrossRef] [PubMed]

38. Winter, J.; Wende, K.; Masur, K.; Iseni, S.; Dunnbier, M.; Hammer, M.U.; Tresp, H.; Weltmann, K.D.; Reuter, S. Feed gas humidity: A vital parameter affecting a cold atmospheric-pressure plasma jet and plasma-treated human skin cells. *J. Phys. D Appl. Phys.* **2013**, *46*, 295401. [CrossRef]

39. Liang, C.C.; Park, A.Y.; Guan, J.L. In vitro scratch assay: A convenient and inexpensive method for analysis of cell migration in vitro. *Nat. Protoc.* **2007**, *2*, 329–333. [CrossRef]

40. Lackmann, J.W.; Wende, K.; Verlackt, C.; Golda, J.; Volzke, J.; Kogelheide, F.; Held, J.; Bekeschus, S.; Bogaerts, A.; Schulz-von der Gathen, V.; et al. Chemical fingerprints of cold physical plasmas—An experimental and computational study using cysteine as tracer compound. *Sci. Rep.* **2018**, *8*, 7736. [CrossRef]

41. England, H.; Summersgill, H.R.; Edye, M.E.; Rothwell, N.J.; Brough, D. Release of interleukin-1α or interleukin-1β depends on mechanism of cell death. *J. Biol. Chem.* **2014**, *289*, 15942–15950. [CrossRef]

42. Martinon, F.; Pétrilli, V.; Mayor, A.; Tardivel, A.; Tschopp, J. Gout-associated uric acid crystals activate the nalp3 inflammasome. *Nature* **2006**, *440*, 237. [CrossRef] [PubMed]

43. Nakae, S.; Asano, M.; Horai, R.; Iwakura, Y. Interleukin-1β, but not interleukin-1α, is required for t-cell-dependent antibody production. *Immunology* **2001**, *104*, 402–409. [CrossRef] [PubMed]

44. Ben-Sasson, S.Z.; Hu-Li, J.; Quiel, J.; Cauchetaux, S.; Ratner, M.; Shapira, I.; Dinarello, C.A.; Paul, W.E. Il-1 acts directly on cd4 t cells to enhance their antigen-driven expansion and differentiation. *Proc. Natl. Acad. Sci. USA* **2009**, *106*, 7119–7124. [CrossRef] [PubMed]

45. O'Sullivan, B.J.; Thomas, H.E.; Pai, S.; Santamaria, P.; Iwakura, Y.; Steptoe, R.J.; Kay, T.W.; Thomas, R. Il-1β breaks tolerance through expansion of cd25+ effector t cells. *J. Immunol.* **2006**, *176*, 7278–7287. [CrossRef] [PubMed]

46. Kapas, L.; Hong, L.; Cady, A.B.; Opp, M.R.; Postlethwaite, A.E.; Seyer, J.M.; Krueger, J.M. Somnogenic, pyrogenic, and anorectic activities of tumor necrosis factor-alpha and tnf-alpha fragments. *Am. J. Physiol.* **1992**, *263*, R708–R715. [CrossRef] [PubMed]

47. Bystry, R.S.; Aluvihare, V.; Welch, K.A.; Kallikourdis, M.; Betz, A.G. B cells and professional apcs recruit regulatory t cells via ccl4. *Nat. Immunol.* **2001**, *2*, 1126–1132. [CrossRef] [PubMed]

48. Bezu, L.; Gomes-de-Silva, L.C.; Dewitte, H.; Breckpot, K.; Fucikova, J.; Spisek, R.; Galluzzi, L.; Kepp, O.; Kroemer, G. Combinatorial strategies for the induction of immunogenic cell death. *Front. Immunol.* **2015**, *6*, 187. [CrossRef]

49. Mizuno, K.; Shirakawa, Y.; Sakamoto, T.; Ishizaki, H.; Nishijima, Y.; Ono, R. Plasma-induced suppression of recurrent and reinoculated melanoma tumors in mice. *IEEE Trans. Radiat. Plasma Med. Sci.* **2018**, *2*, 353–359. [CrossRef]

50. Mizuno, K.; Yonetamari, K.; Shirakawa, Y.; Akiyama, T.; Ono, R. Anti-tumor immune response induced by nanosecond pulsed streamer discharge in mice. *J. Phys. D Appl. Phys.* **2017**, *50*, 12LT01. [CrossRef]

51. Bekeschus, S.; Kolata, J.; Muller, A.; Kramer, A.; Weltmann, K.-D.; Broker, B.; Masur, K. Differential viability of eight human blood mononuclear cell subpopulations after plasma treatment. *Plasma Med.* **2013**, *3*, 1–13. [CrossRef]

52. Bundscherer, L.; Bekeschus, S.; Tresp, H.; Hasse, S.; Reuter, S.; Weltmann, K.-D.; Lindequist, U.; Masur, K. Viability of human blood leukocytes compared with their respective cell lines after plasma treatment. *Plasma Med.* **2013**, *3*, 71–80. [CrossRef]

53. Bekeschus, S.; Rödder, K.; Schmidt, A.; Stope, M.B.; von Woedtke, T.; Miller, V.; Fridman, A.; Weltmann, K.-D.; Masur, K.; Metelmann, H.-R.; et al. Cold physical plasma selects for specific t helper cell subsets with distinct cells surface markers in a caspase-dependent and nf-κb-independent manner. *Plasma Process. Polym.* **2016**, *13*, 1144–1150. [CrossRef]

54. Saito, Y.; Nishio, K.; Ogawa, Y.; Kimata, J.; Kinumi, T.; Yoshida, Y.; Noguchi, N.; Niki, E. Turning point in apoptosis/necrosis induced by hydrogen peroxide. *Free Radic. Res.* **2006**, *40*, 619–630. [CrossRef] [PubMed]

55. Reth, M. Hydrogen peroxide as second messenger in lymphocyte activation. *Nat. Immunol.* **2002**, *3*, 1129. [CrossRef] [PubMed]

56. Premack, B.A.; Gardner, P. Signal transduction by t-cell receptors: Mobilization of ca and regulation of ca-dependent effector molecules. *Am. J. Physiol.-Cell Physiol.* **1992**, *263*, C1119–C1140. [CrossRef] [PubMed]

57. Lewis, R.S.; Cahalan, M.D. Potassium and calcium channels in lymphocytes. *Annu. Rev. Immunol.* **1995**, *13*, 623–653. [CrossRef] [PubMed]

58. Goldsmith, M.A.; Weiss, A. Early signal transduction by the antigen receptor without commitment to t cell activation. *Science* **1988**, *240*, 1029–1031. [CrossRef] [PubMed]

59. Atherfold, P.; Norris, M.; Robinson, P.; Gelfand, E.; Franklin, R. Calcium-induced erk activation in human t lymphocytes. *Mol. Immunol.* **1999**, *36*, 543–549. [CrossRef]

60. Franklin, R.A.; Atherfold, P.A.; McCubrey, J.A. Calcium-induced erk activation in human t lymphocytes occurs via p56lck and cam-kinase. *Mol. Immunol.* **2000**, *37*, 675–683. [CrossRef]

61. Bekeschus, S.; Masur, K.; Kolata, J.; Wende, K.; Schmidt, A.; Bundscherer, L.; Barton, A.; Kramer, A.; Broker, B.; Weltmann, K.D. Human mononuclear cell survival and proliferation is modulated by cold atmospheric plasma jet. *Plasma Process. Polym.* **2013**, *10*, 706–713. [CrossRef]

62. Van der Linde, J.; Liedtke, K.R.; Matthes, R.; Kramer, A.; Heidecke, C.-D.; Partecke, L.I. Repeated cold atmospheric plasma application to intact skin does not cause sensitization in a standardized murine model. *Plasma Med.* **2017**, *7*, 383–393. [CrossRef]

63. Brune, B.; Dehne, N.; Grossmann, N.; Jung, M.; Namgaladze, D.; Schmid, T.; von Knethen, A.; Weigert, A. Redox control of inflammation in macrophages. *Antioxid. Redox Signal.* **2013**, *19*, 595–637. [CrossRef] [PubMed]

64. Partida-Sanchez, S.; Iribarren, P.; Moreno-Garcia, M.E.; Gao, J.L.; Murphy, P.M.; Oppenheimer, N.; Wang, J.M.; Lund, F.E. Chemotaxis and calcium responses of phagocytes to formyl peptide receptor ligands is differentially regulated by cyclic adp ribose. *J. Immunol.* **2004**, *172*, 1896–1906. [CrossRef] [PubMed]

65. Notter, M.; Schirrmacher, V. Tumor-specific t-cell clones recognize different protein determinants of autologous human malignant melanoma cells. *Int. J. Cancer* **1990**, *45*, 834–841. [CrossRef]

66. Anichini, A.; Mazzocchi, A.; Fossati, G.; Parmiani, G. Cytotoxic t lymphocyte clones from peripheral blood and from tumor site detect intratumor heterogeneity of melanoma cells. Analysis of specificity and mechanisms of interaction. *J. Immunol.* **1989**, *142*, 3692–3701. [PubMed]

67. Wolfel, T.; Hauer, M.; Klehmann, E.; Brichard, V.; Ackermann, B.; Knuth, A.; Boon, T.; Meyer Zum Buschenfelde, K.H. Analysis of antigens recognized on human melanoma cells by a2-restricted cytolytic t lymphocytes (ctl). *Int. J. Cancer* **1993**, *55*, 237–244. [CrossRef] [PubMed]

68. Shahinian, A.; Pfeffer, K.; Lee, K.P.; Kundig, T.M.; Kishihara, K.; Wakeham, A.; Kawai, K.; Ohashi, P.S.; Thompson, C.B.; Mak, T.W. Differential t cell costimulatory requirements in cd28-deficient mice. *Science* **1993**, *261*, 609–612. [CrossRef] [PubMed]

69. Linsley, P.S.; Ledbetter, J.A. The role of the cd28 receptor during t cell responses to antigen. *Annu. Rev. Immunol.* **1993**, *11*, 191–212. [CrossRef]

70. Vella, A.T.; Mitchell, T.; Groth, B.; Linsley, P.S.; Green, J.M.; Thompson, C.B.; Kappler, J.W.; Marrack, P. Cd28 engagement and proinflammatory cytokines contribute to t cell expansion and long-term survival in vivo. *J. Immunol.* **1997**, *158*, 4714–4720. [PubMed]

71. Sperling, A.I.; Auger, J.A.; Ehst, B.D.; Rulifson, I.C.; Thompson, C.B.; Bluestone, J.A. Cd28/b7 interactions deliver a unique signal to naive t cells that regulates cell survival but not early proliferation. *J. Immunol.* **1996**, *157*, 3909–3917. [PubMed]

72. Mueller, D.L.; Seiffert, S.; Fang, W.; Behrens, T.W. Differential regulation of bcl-2 and bcl-x by cd3, cd28, and the il-2 receptor in cloned cd4+ helper t cells. A model for the long-term survival of memory cells. *J. Immunol.* **1996**, *156*, 1764–1771. [PubMed]

73. Arosa, F.A. Cd8+cd28- t cells: Certainties and uncertainties of a prevalent human t-cell subset. *Immunol. Cell Biol.* **2002**, *80*, 1–13. [CrossRef] [PubMed]

74. Berner, B.; Akca, D.; Jung, T.; Muller, G.A.; Reuss-Borst, M.A. Analysis of th1 and th2 cytokines expressing cd4+ and cd8+ t cells in rheumatoid arthritis by flow cytometry. *J. Rheumatol.* **2000**, *27*, 1128–1135. [PubMed]

75. Morelli, A.E. Cytokine production by mouse myeloid dendritic cells in relation to differentiation and terminal maturation induced by lipopolysaccharide or cd40 ligation. *Blood* **2001**, *98*, 1512–1523. [CrossRef] [PubMed]

76. Fleischer, J.; Soeth, E.; Reiling, N.; GRAGE-GRIEBENOW, E.; FLAD, H.D.; Ernst, M. Differential expression and function of cd80 (b7-1) and cd86 (b7-2) on human peripheral blood monocytes. *Immunology* **1996**, *89*, 592–598. [CrossRef] [PubMed]

77. Girndt, M.; Sester, M.; Sester, U.; Kaul, H.; Köhler, H. Defective expression of b7-2 (cd86) on monocytes of dialysis patients correlates to the uremia-associated immune defect. *Kidney Int.* **2001**, *59*, 1382–1389. [CrossRef] [PubMed]

78. Maurer, D.; Fiebiger, E.; Reininger, B.; Wolff-Winiski, B.; Jouvin, M.-H.; Kilgus, O.; Kinet, J.-P.; Stingl, G. Expression of functional high affinity immunoglobulin e receptors (fc epsilon ri) on monocytes of atopic individuals. *J. Exp. Med.* **1994**, *179*, 745–750. [CrossRef] [PubMed]

79. Melewicz, F.; Zeiger, R.; Mellon, M.; O'Connor, R.; Spiegelberg, H. Increased ige-dependent cytotoxicity by blood mononuclear cells of allergic patients. *Clin. Exp. Immunol.* **1981**, *43*, 526.

80. Nucera, S.; Biziato, D.; De Palma, M. The interplay between macrophages and angiogenesis in development, tissue injury and regeneration. *Int. J. Dev. Biol.* **2011**, *55*, 495–503. [CrossRef]

81. Italiani, P.; Boraschi, D. From monocytes to m1/m2 macrophages: Phenotypical vs. Functional differentiation. *Front. Immunol.* **2014**, *5*, 514. [CrossRef] [PubMed]

82. Bekeschus, S.; Freund, E.; Wende, K.; Gandhirajan, R.; Schmidt, A. Hmox1 upregulation is a mutual marker in human tumor cells exposed to physical plasma-derived oxidants. *Antioxidants* **2018**, *7*, 151. [CrossRef] [PubMed]

83. Bekeschus, S.; Kading, A.; Schroder, T.; Wende, K.; Hackbarth, C.; Liedtke, K.R.; van der Linde, J.; von Woedtke, T.; Heidecke, C.D.; Partecke, L.I. Cold physical plasma treated buffered saline solution as effective agent against pancreatic cancer cells. *Anticancer Agents Med. Chem.* **2018**, *18*, 824–831. [CrossRef] [PubMed]

84. Bekeschus, S.; Lin, A.; Fridman, A.; Wende, K.; Weltmann, K.D.; Miller, V. A comparison of floating-electrode dbd and kinpen jet: Plasma parameters to achieve similar growth reduction in colon cancer cells under standardized conditions. *Plasma Chem. Plasma Process.* **2018**, *38*, 1–12. [CrossRef]

85. Bekeschus, S.; Mueller, A.; Miller, V.; Gaipl, U.; Weltmann, K.-D. Physical plasma elicits immunogenic cancer cell death and mitochondrial singlet oxygen. *IEEE Trans. Radiat. Plasma Med. Sci.* **2018**, *2*, 138–146. [CrossRef]

86. Bekeschus, S.; Wende, K.; Hefny, M.M.; Rodder, K.; Jablonowski, H.; Schmidt, A.; Woedtke, T.V.; Weltmann, K.D.; Benedikt, J. Oxygen atoms are critical in rendering thp-1 leukaemia cells susceptible to cold physical plasma-induced apoptosis. *Sci. Rep.* **2017**, *7*, 2791. [CrossRef] [PubMed]

87. Bekeschus, S.; Wulf, C.; Freund, E.; Koensgen, D.; Mustea, A.; Weltmann, K.-D.; Stope, M. Plasma treatment of ovarian cancer cells mitigates their immuno-modulatory products active on thp-1 monocytes. *Plasma* **2018**, *1*, 201–217. [CrossRef]

88. Adachi, T.; Tanaka, H.; Nonomura, S.; Hara, H.; Kondo, S.; Hori, M. Plasma-activated medium induces a549 cell injury via a spiral apoptotic cascade involving the mitochondrial-nuclear network. *Free Radic. Biol. Med.* **2015**, *79*, 28–44. [CrossRef]

89. Ahn, H.J.; Kim, K.I.; Hoan, N.N.; Kim, C.H.; Moon, E.; Choi, K.S.; Yang, S.S.; Lee, J.S. Targeting cancer cells with reactive oxygen and nitrogen species generated by atmospheric-pressure air plasma. *PLoS ONE* **2014**, *9*, e86173. [CrossRef]

90. Akhlaghi, M.; Rajaei, H.; Mashayekh, A.S.; Shafiae, M.; Mahdikia, H.; Khani, M.; Hassan, Z.M.; Shokri, B. Determination of the optimum conditions for lung cancer cells treatment using cold atmospheric plasma. *Phys. Plasmas* **2016**, *23*. [CrossRef]

91. Aryal, S.; Bisht, G. New paradigm for a targeted cancer therapeutic approach: A short review on potential synergy of gold nanoparticles and cold atmospheric plasma. *Biomedicines* **2017**, *5*, 38. [CrossRef]

92. Boehm, D.; Curtin, J.; Cullen, P.J.; Bourke, P. Hydrogen peroxide and beyond-the potential of high-voltage plasma-activated liquids against cancerous cells. *Anticancer Agents Med. Chem.* **2018**, *18*, 815–823. [CrossRef] [PubMed]

93. Chang, J.W.; Kang, S.U.; Shin, Y.S.; Kim, K.I.; Seo, S.J.; Yang, S.S.; Lee, J.S.; Moon, E.; Lee, K.; Kim, C.H. Non-thermal atmospheric pressure plasma inhibits thyroid papillary cancer cell invasion via cytoskeletal modulation, altered mmp-2/-9/upa activity. *PLoS ONE* **2014**, *9*, e92198. [CrossRef] [PubMed]

94. Chang, J.W.; Kang, S.U.; Shin, Y.S.; Kim, K.I.; Seo, S.J.; Yang, S.S.; Lee, J.-S.; Moon, E.; Baek, S.J.; Lee, K. Non-thermal atmospheric pressure plasma induces apoptosis in oral cavity squamous cell carcinoma: Involvement of DNA-damage-triggering sub-g_1 arrest via the atm/p53 pathway. *Arch. Biochem. Biophys.* **2014**, *545*, 133–140. [CrossRef]

95. Chen, Z.; Lin, L.; Cheng, X.; Gjika, E.; Keidar, M. Treatment of gastric cancer cells with nonthermal atmospheric plasma generated in water. *Biointerphases* **2016**, *11*, 031010. [CrossRef] [PubMed]

96. Chen, Z.; Simonyan, H.; Cheng, X.; Gjika, E.; Lin, L.; Canady, J.; Sherman, J.H.; Young, C.; Keidar, M. A novel micro cold atmospheric plasma device for glioblastoma both in vitro and in vivo. *Cancers* **2017**, *9*, 61. [CrossRef] [PubMed]

97. Chen, Z.T.; Lin, L.; Cheng, X.Q.; Gjika, E.; Keidar, M. Effects of cold atmospheric plasma generated in deionized water in cell cancer therapy. *Plasma Process. Polym.* **2016**, *13*, 1151–1156. [CrossRef]

98. Cheng, X.; Sherman, J.; Murphy, W.; Ratovitski, E.; Canady, J.; Keidar, M. The effect of tuning cold plasma composition on glioblastoma cell viability. *PLoS ONE* **2014**, *9*, e98652. [CrossRef]

99. Kumar, N.; Attri, P.; Dewilde, S.; Bogaerts, A. Inactivation of human pancreatic ductal adenocarcinoma with atmospheric plasma treated media and water: A comparative study. *J. Phys. D Appl. Phys.* **2018**, *51*. [CrossRef]

100. Kumar, N.; Park, J.H.; Jeon, S.N.; Park, B.S.; Choi, E.H.; Attri, P. The action of microsecond-pulsed plasma-activated media on the inactivation of human lung cancer cells. *J. Phys. D Appl. Phys.* **2016**, *49*. [CrossRef]

101. Schmidt, A.; Bekeschus, S.; Jarick, K.; Hasse, S.; von Woedtke, T.; Wende, K. Cold physical plasma modulates p53 and mitogen-activated protein kinase signaling in keratinocytes. *Oxid. Med. Cell. Longev.* **2019**, *2019*, 1–16. [CrossRef]

 *applied sciences*

Review

Application of Non-Thermal Plasma on Biofilm: A Review

Tripti Thapa Gupta [1,*,†] **and Halim Ayan** [1,2]

[1] Department of Bioengineering, College of Engineering, University of Toledo, Toledo, OH 43606, USA
[2] Department of Mechanical, Industrial and Manufacturing Engineering, College of Engineering, University of Toledo, Toledo, OH 43606, USA
* Correspondence: tripti.thapa@utoledo.edu
† Department of Microbial Infection and Immunity, College of Medicine, Ohio State University, Columbus, OH 43210, USA.

Received: 9 July 2019; Accepted: 27 August 2019; Published: 29 August 2019

Abstract: The formation of bacterial biofilm on implanted devices or damaged tissues leads to biomaterial-associated infections often resulting in life-threatening diseases and implant failure. It is a challenging process to eradicate biofilms as they are resistant to antimicrobial treatments. Conventional techniques, such as high heat and chemicals exposure, may not be suitable for biofilm removal in nosocomial settings. These techniques create surface degradation on the treated materials and lead to environmental pollution due to the use of toxic chemicals. A novel technique known as non-thermal plasma has a great potential to decontaminate or sterilize those nosocomial biofilms. This article aims to provide readers with an extensive review of non-thermal plasma and biofilms to facilitate further investigations. A brief introduction summarizes the problem caused by biofilms in hospital settings with current techniques used for biofilm inactivation followed by the literature review strategy. The remainder of the review discusses plasma and its generation, the role played by plasma reactive species, various factors affecting the antimicrobial efficacy of non-thermal plasma and summarizes many studies published in the field.

Keywords: biofilm; decontamination; dielectric barrier discharge; infection; jet plasma; non-thermal plasma

---

1. Introduction

Biofilms refer to a group of microorganisms adhered to a substrate within a polymeric matrix [1]. Biofilms possess unfavorable conditions like biofouling, pipe plugging, damage of equipment, prosthesis colonization, and a number of diseases [2]. Due to their greater resistance to antimicrobial treatment than planktonic cells of the same species, biofilms are usually challenging to eradicate [3–5]. Approximately 13 million people suffer from biofilm-related infections in the United States [6]. The Centers for Diseases Control and Prevention (CDC) estimates that approximately 65% of human bacterial diseases are due to biofilms, with a higher estimate (80%) proposed by the National Institutes of Health (NIH) [7]. More than 60% of hospital-acquired infections (HAI) are led by the attachment of a number of microorganisms to medical implants/devices like catheters, prostheses, fracture-fixation devices, dental implants, and cardiac devices [5,8–14]. In Europe and USA, the World Health Organization estimates that about 4.5 million and 1.7 million patients are impacted by healthcare associated infections that lead to 100,000 and 37,000 deaths per annum [15]. Most of the infections that occur in nosocomial settings are due to the growth of bacteria such as *Escherichia coli*, *Pseudomonas aeruginosa*, *Staphylococcus aureus*, *Streptococcus pyrogens*, and *Candida albicans* in biofilm form [16].

Critical requirements for developing an antimicrobial tool against biofilms in hospital settings are robustness and the disinfection and sterilization ability without any human side effects or damage to the medical devices and implants. Traditional sterilization methods such as heat and chemical treatments (quaternary ammonium compounds, aldehydes, alcohols and halogens or radiation, chlorhexidine and silver salts, peroxygens, glutaraldehyde, and ortho-phthalaldehyde) [8,17,18] cannot be considered as perfect bacterial decontamination techniques because of their potential to eventually degrade the treated surface. Moreover, chemical usage can be toxic and lead to environmental pollution. As biofilms are significantly more resistant to antimicrobial treatment than in the planktonic form of the same species [3], removing them by conventional treatments would be problematic. Also, antibiotic treatment may not always inactivate overall bacterial cells present in the biofilm as they are more resistant to antibiotics. In addition to their limited efficacy, other drawbacks of current sterilization and disinfection methods include their environmental impact, clinical downtime and economic costs [19]. Therefore, a novel antimicrobial treatment technique, such as a non-thermal plasma (NTP), one which is safer, efficient, and cost-effective in clinical settings, will be of great interest.

The main scope of this paper is to provide a review of the published research in the area of biofilm decontamination and sterilization using plasma. The emphasis here is on the potential of NTP for biofilm inactivation. The basics of this novel technology is reviewed, and a brief overview is given of extensive published research in the field. Other sections provide explanations and reviews of biofilms and their formation, the varieties of NTP, plasma generation, various plasma reactive species, factors influencing the antimicrobial efficacy of plasma, and summaries of several significant publications in the field.

2. Literature Review Strategy

This review was prepared through an extensive literature survey. PubMed and Google Scholar were used to search for literature, with no date restriction in the field of NTP and biofilm. The following keywords and phrases were used to find relevant published articles: "medical biofilm", "hospital infection by biofilm", "biofilm formation", "dielectric barrier discharge", "NTP and biofilm", "NTP towards biofilm decontamination/sterilization", "antimicrobial efficacy of NTP". Since the articles were of heterogeneous nature, it was impossible to apply rigid selection criteria. Instead, articles were chosen based on their biofilm focus, including investigations of NTP usage for biofilm sterilization, or decontamination. In addition, the reference lists of each article found during the primary search were reviewed to identify other relevant literature.

3. Biofilm and Its Formation

Understanding the characteristics and formation of biofilms would be the first step towards their decontamination or sterilization via NTP. Knowing the structure of biofilms is important as it serves to protect them against several antibiotics and sterilization procedures. A biofilm refers to a group of microorganisms adhered to biotic or abiotic surfaces. These adherent microorganisms are enclosed within a self-produced matrix of extracellular polymeric substance (EPS). The biofilm matrix is composed of 97% water and mainly polysaccharides, proteins, and extracellular DNA (eDNA) which provide the biofilm structure and also act as a reservoir for nutrients [20]. Out of the total organic materials present in the biofilm, 75–90% is EPS, 10–25% is microbial cells and 1–2% is proteins, polysaccharides, peptidoglycans, lipids, phospholipids, DNA, and RNA [21]. Biofilms are present in diverse environments such as in households, water sources, pipes, industry, and hospitals. Their presence in hospitals can lead to diseases, prostheses colonization, product contamination, biofouling, and equipment damage [22]. Various bacterial species such as *Pseudomonas*, *Staphylococcus* and *Candida* are frequently isolated in clinically relevant biofilm form [2]. Commonly isolated microorganisms from indwelling medical devices are shown in Table 1.

Table 1. Biofilm-associated microorganisms commonly isolated from selected indwelling medical devices [23].

Indwelling Medical Device	Organisms
Central venous catheter	Coagulase-negative staphylococci, *Staphylococcus aureus*, *Enterococcus faecalis*, *Klebsiella pneumoniae*, *Pseudomonas aeruginosa*, *Candida albicans*
Prosthetic heart valve	Viridans *Streptococcus*, coagulase-negative staphylococci, enterococci, *Staphylococcus aureus*
Urinary catheter	*Staphylococcus epidermidis*, *Escherichia coli*, *Klebsiella pneumoniae*, *Enterococcus faecalis*, *Proteus mirabilis*
Artificial hip prosthesis	Coagulase-negative staphylococci, b-hemolytic streptococci, enterococci, *Proteus mirabilis*, *Bacterioides* species, *Staphylococcus aureus*, viridans *Streptococcus*, *Escherichia coli*, *Pseudomonas aeruginosa*
Artificial voice prosthesis	*Candida albicans*, *Streptococcus mitis*, *Streptococcus salivarius*, *Rothia dentrocariosa*, *Candida tropicalis*, *Streptococcus sobrinus*, *Staphylococcus epidermidis*, Stomatococcus mucilaginous
Intrauterine device	*Staphylococcus epidermidis*, *Corynebacterium* species, *Staphylococcus aureus*, *Micrococcus* species, *Lactobacillus plantarum*, Group B streptococci, *Enterococcus species*, *Candida albicans*

The biofilm formation is initiated after free floating planktonic bacteria attach on a substrate. Within a few hours of attachment, these bacteria bind irreversibly and begin to multiply resulting microcolonies on the surface and produce a polymeric substance around them as shown in Figure 1. This biofilm when matured becomes resistant to antibiotics. Production of EPS is the maturation stage of the biofilm that bind cells together on the surface [24]. EPS also forms a physical barrier which is responsible for limiting the transport of chemicals into and out of the biofilm [24].

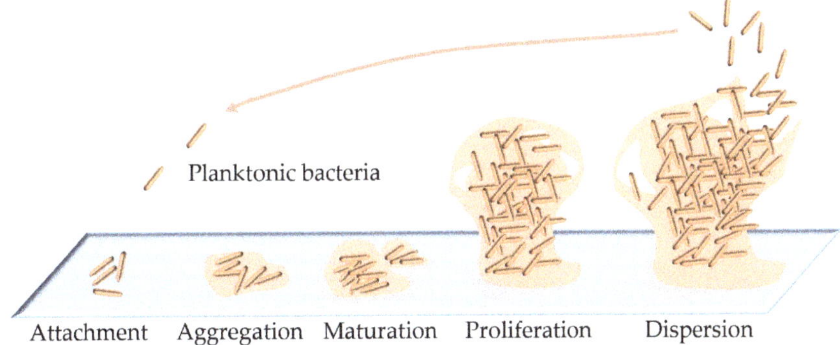

Figure 1. The life cycle of biofilms beginning with the initial cell attachment and ending with biofilm dispersal.

Nutrient depletion, metabolic product accumulation, and other stressors cause cells from the biofilm to disperse. Ultimately, the dispersion of biofilm refers to the final stage of its life cycle and consists of three phases—detachment from the existing biofilm, translocation to other areas, and adherence/colonization on other surfaces [25]. Dispersion of the biofilm spreads the infection to other areas and increases the contamination of a medical device [21]. Enzymes that are responsible for degrading the extracellular matrix, for example, dispersin B and deoxyribonuclease may play roles in this dispersal [26].

Quorum sensing (QS) is a communicative system for biofilm which allows the bacterial cells inside the biofilm to act as a community. This facilitates the growth, survival, and colonization of bacterial cells inside the biofilm [21]. The concentration of other bacteria present within a limited microenvironment is sensed through QS [27]. The bacteria then respond by activating specific genes which then produce virulence factors such as enzymes or toxins [27]. The QS molecules are N-acyl-L-homoserine lactones (AHL) in gram-negative bacteria and peptides in gram-positive bacteria [27].

Many investigations of the mechanism of biofilm resistance to antimicrobial agents have been carried out. A primary reason is the biofilm matrix preventing the access of antimicrobial agents, such as antibiotics, to the embedded bacterial cells [28]. Other reasons include slow penetration, resistant phenotypes, and altered microenvironments [29]. Because of slow penetration, antibiotics may fail to penetrate deeply into the biofilm. Some of the bacteria in the biofilm may develop into a protected phenotype resistant to antibiotics. Similarly, the accumulation of waste and nutrient depletion in the deeper biofilm layers may contribute to antibiotic resistance.

4. Non-Thermal Plasma

Non-thermal plasma (NTP) is a novel and emerging antimicrobial tool that demonstrates the possibility of improved biofilm decontamination and sterilization [30–33]. Plasma is the fourth fundamental states of matter after solid, liquid, and gas. The various ingredients present in plasma such as free radicals, reactive oxygen, and nitrogen species [34], and positive and negative ions [35,36] are believed to play a significant role as antimicrobial agents. Thermal (equilibrium) and non-thermal (non-equilibrium) plasma are the two plasma types. They are differentiated based on the relative energy levels of electrons and heavy particles of the plasma [37]. A thermal plasma is generated at high pressure and power and contains electrons and heavy particles at the same temperature. In contrast, NTP is generated at low pressure and power, and contains electrons at higher temperature and heavy particles at room temperature [37,38].

Previously, thermal plasma was used for tissue removal, sterilization, and cauterization [39]. The problem with this type of plasma treatment is the high heat production and damage to tissues and other surfaces. Conversely, NTP can carry out the same functions without causing any such harm or side effects [40]. NTPs such as Dielectric Barrier Discharge (DBD) and jet plasmas are gaining much interest because of their non-thermal nature. It enables new applications in biological and medical fields where substratum to be treated are mostly living tissue, cells, and biomaterials [41]. As listed in Table 2, recent investigations with NTP show promising results for the sterilization and decontamination of biofilms of different bacterial species with this novel technology.

Table 2. Uses of non-thermal plasma (NTP) for biofilm decontamination or sterilization.

Bacterial Strain	Plasma Type and Parameters (Voltage, Frequency, Power, Working Gas, and Flow Rate)	Inactivation Yield	Substrate for Biofilm Formation with Time for Decontamination or Sterilization	Ref.
Neisseria gonorrhoeae	Jet: 10 kV and 10 kHz, He at 2 L/min	7 log reduction	Coverslips (20 min)	[42]
Enterococcus faecalis	Jet: 18 kV and 10 kHz, Ar/O$_2$ (2%) at 5 L/min	No CFU detected	Root canal (10 min)	[43]
Streptococcus mutants	DBD: 580 kHz and 2 W/cm^3 power density, He at 2 L/min	98% killed	Tooth slices (30 s)	[44]

Table 2. *Cont.*

Bacterial Strain	Plasma Type and Parameters (Voltage, Frequency, Power, Working Gas, and Flow Rate)	Inactivation Yield	Substrate for Biofilm Formation with Time for Decontamination or Sterilization	Ref.
Weissella confusa	Jet: 20 kHz	2.63 and 2.16 log reduction with and without sucrose	Cellulose ester membrane (20 min)	[45]
Porphyromonas gingivalis	Jet: 8 kV and 8 kHz, He/O_2 (1%) at 1 L/min	All cells killed in 15 µm biofilm	Cover slip (5 min)	[46]
Staphylococcus aureus	FE-DBD: 120 V and 0.13 W/cm^2 power	All biofilms were sterilized	Cover slip and 96 well plate (<2 min)	[47]
Pseudomonas aeruginosa	Jet: 6 kV with 20 and 40 kHz, He/O_2 (0.5%) at 2 L/min	4 log reduction at 20 kHz and complete eradication at 40 kHz	Peg lid of Calgary biofilm device and polycarbonate coupon (4 min)	[3]
Pseudomonas aeruginosa	DBD: 120 kV and 50 Hz	Biofilm reduced to undetectable level	96 well plate and coverslips (5 min)	[48]
Staphylococcus aureus	Jet: 20 kV and 38 kHz, He at 6.7 L/min	3.06 log reduction	Borosilicate slices (10 min)	[49]
Staphylococcus aureus, epidermidis and *Escherichia coli*	Low power gas discharge: 60 W power, oxygen, argon, and nitrogen at flow rate of 2.4 ft^3 h^{-1}	All biofilms killed	Polyethylene terephthalate (PET) films, silicon wafers and cover-glass chambers (25–30 min)	[50]
Burkholderia cenocepacia and Pseudomonas aeruginosa	MicroPlaSter B device plasma with argon gas	0.005% and 2% bacteria survived	Coverglass (10 min)	[51]
Staphylococcus aureus, Pseudomonas aeruginosa and *Candida albicans*	Jet: 5 kV and 61 kHz, 2.5–3.5 W power, Ar+O_2	27%, 39%, and 35% cells survived of *S. aureus*, *P. aeruginosa*, and *C. albicans*	96 well plate (3 min)	[52]
Streptococcus mutans and saliva multispecies	Jet (KINPen 09): Ar (5 slm) and Ar+1%(O_2) at 0.05 slm, HDBD (hollow DBD): 8.4 kV and 37.6 kHz, Ar and Ar+O_2 at 1 and 0.01 slm, and VDBD (volume DBD):10 kV and 40 kHz, Ar at 0.05 slm	5.38 for *S. mutans* and 5.67 for saliva biofilm	Titanium disc (10 min)	[31]
Pseudomonas aeruginosa and *Staphylococcus epidermidis*	Surface dielectric barrier discharge (SBD)-Structured electrode planar SBDA: (13 kV and 20 kHz) and a wire electrode SBD-B: (8 kV and 30 kHz with compressed air at 0.5 slm)	7.1 and 3.8 log reduction by SBD-A in *P. aeruginosa* and SBD-B and 3.4 and 2.7 log reduction by SBD-A and SBD-B in *S. epidermidis*	Polycarbonate disc (10 min)	[53]

Table 2. *Cont.*

Bacterial Strain	Plasma Type and Parameters (Voltage, Frequency, Power, Working Gas, and Flow Rate)	Inactivation Yield	Substrate for Biofilm Formation with Time for Decontamination or Sterilization	Ref.
Candida albicans	Jet(KINPen09): 220 V and 50/60 Hz, 8 W power, Ar at 5 slm and Ar+(1%)O_2 at 0.05 slm, HDBD:9 kV and 37.6 kHz, 9W power, Ar at 6 slm and Ar+(1%)O_2 at 0.06 slm and VDBD: 10 kV and 40 kHz, 16 W power	5 log reduction	Titanium disc (10 min)	[54]
Candida albicans	Surface microdischarge plasma technology (SMD): 9 kV and 1 kHz	6 log reduction	6 well plate (8 min)	[55]
Pseudomonas aeruginosa	Atomflo 300 reactor plasma Jet: 13.56 MHz and 100 W R, He at 20.4 L/min and N_2 at 0.135 L/min at 35 W	Complete biofilm inactivation	Borosilicate coupon (30 min)	[56]
Pseudomonas aeruginosa and *Staphylococcus epidermidis*	Jet (Kinpen09): 2–6 kVpp and 1.1 MHz, 3.5 W power, Ar and Ar+(1%)O_2 at 5 slm	5.41 and 5.10 log reduction in Ar and Ar + O_2 plasma for *P. aeruginosa* and 3.14 and 2.21 log reduction in Ar and Ar + O_2 plasma for *S. epidermidis*	Microtiter plate (5 min)	[57]
Enterococcus faecalis	Plasma dental probe:6 kV and 1 kHz, 0.7 W power, He/(1%)O_2 at 1 slm	93.1% biofilm killing	Hydroxyapatite discs (5 min)	[58]
Bacillus cereus, Staphylococcus aureus, Escherichia coli and *Pseudomonas aeruginosa*	Jet:6 kV and 20 kHz, He/(0.5%)O_2 at 2 slm	Complete biofilm eradication	Peg lid of Calgary Biofilm Device (<4 min and 10 min)	[59]
Streptococci	Atmospheric pressure air plasma of corona discharge: Positive corona (PC) −8 kV and 20 kHz, negative corona (NC) −7 kV and 0.25 to 2 MHz	3 log reduction	Tooth surfaces (10 min)	[60]
Candida albicans	Jet (KINPen 08): 2–6 kVpp and 1.7 MHz, Ar at 5 slm and Ar/[1%]O_2 at 0.05 slm	Complete biofilm removal with Ar+O_2	Polystyrene (PS) wafers (5 min)	[61]
Pseudomonas aeruginosa	Atomflo 300 reactor plasma Jet: 13.56 MHz and 100 W RF, He at 20.4 L/min and N_2 at 0.135 L/min at 35 W	100% ianctivation	CDC biofilm reactor on borosilicate coupons (5 min)	[62]

Non-thermal jet plasma devices employing atmospheric pressure plasma are commercially available [63]. kINPen® is a pen-like device developed for biomedical applications that allows precise and arbitrary movements in 3D [64]. The plasma is generated after applying a high-frequency (HF) voltage coupled to the pin-type electrode [63]. It is electrically safe to use as it is certified and complies with EU standards [64]. The feed gas used is argon with the capability of using other gases in smaller amounts [65]. The kINPen® MED (Leibniz Institute for Plasma Science and Technology-INP Greifswald and neoplas tools GmbH, Greifswald, Germany) is a predecessor device of KINPen® 09 and is the first atmospheric pressure plasma jet device to be accredited as a medical device (Class IIa) for patient use [64]. However, both kINPen® 09 and kINPen® MED are essentially similar [64]. kINPen® plasma has been clinically used for antimicrobial efficacy and for wound healing in animal and clinical observational studies [64]. To our knowledge, only three other plasma sources have been accredited as medical devices. These are SteriPlas (AdTec Ltd., Japan), PlasmaDerm (Cinogy GmbH, Duderstadt, Germany), and PlasmaOne (Medical Systems GmbH, Bad Ems, Germany) [64]. These plasma sources and several others have been used in various biomedical applications such as wound healing and chronic leg ulcers [66,67], reducing bacterial populations in wounds [68], and biofilm decontamination or sterilization [69,70].

Plasma has enormous potential in several biomedical research areas including sterilization of implant surfaces, surface modification [71], in-vitro blood coagulation [72,73], wound healing and disinfection [74,75], tissue regeneration [39,73], treatment of various infections [39], bacterial decontamination and sterilization [69,76,77], dental cavities [78,79], and cancer treatment [80–82]. The methods used for plasma generation are DBD, atmospheric pressure plasma jet (APPJ), plasma needle, and plasma pencil [83]. These plasmas can be produced in air or with various gases, such as oxygen, helium, argon, and nitrogen. Plasma can be produced by power sources with different frequencies such as low frequency, radio frequency, microwave frequency, high voltage AC or DC, to generate atmospheric and low-pressure glow discharge, corona, magnetron, microwave, gliding arc, plasma jet, and DBD discharge [84–86].

DBD has been known for more than a century. Floating-electrode dielectric barrier discharge (FE-DBD) is a DBD-based plasma source that is regarded as the starting point of modern plasma medicine [87]. FE-DBD is an electrical discharge between two electrodes at atmospheric pressure and air where one electrode is grounded, and the other is supplied with a high voltage (Figure 2). The high voltage electrode is enclosed within a dielectric material that limits the discharge current and the formation of an arc. Different dielectric materials used are quartz, glass or silica glass, polymers, ceramics, thin enamel, or polymer layers that block DC current [88]. The distance between the two electrodes ranges from micrometers to centimeters depending on the operating voltage, process gas and the plasma configuration employed [89]. DBD are usually operated at frequencies up to several tens of kHz. Two DBD configurations that have been employed for most of the applications are parallel and concentric configurations. The operating principle of DBD is shown in Figure 3a–d below. Three different types of DBDs are filamentary, patterned, and diffuse DBD [90]. These DBD types depend upon the construction and operating conditions of the plasma sources.

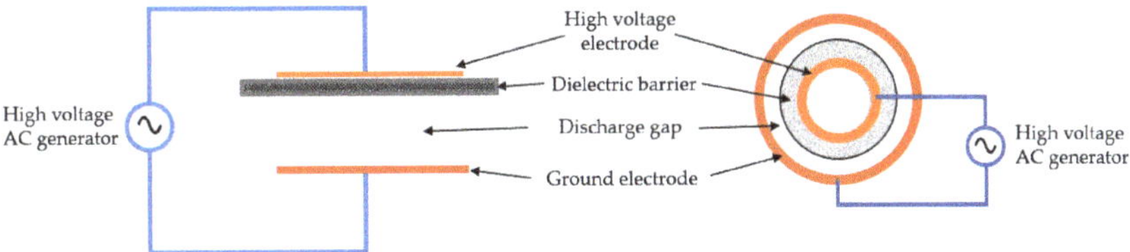

Figure 2. Parallel and concentric dielectric barrier discharge (DBD) configurations [91].

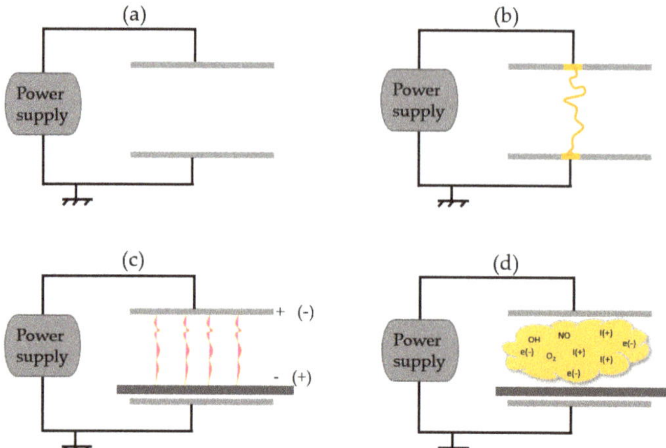

Figure 3. Schematics of DBD plasma operation. (**a,b**) arc production between two electrodes without insulation. (**c,d**) NTP generation after placement of dielectric barrier layer in one of the electrode [41].

Plasma jet is another type of NTP configuration. It is a tube with electrodes which are either placed inside or around it, through which gas flows and ionizes when it is subjected to the electrical field between the electrodes [92]. Figures 4b and 5b show the pictorial representations of volume DBD and jet plasma. In volume DBD, plasma is ignited in the gap between the high voltage electrode and the sample that is connected to the ground electrode. Whereas in the jet configuration, plasma is ignited inside the nozzle/tube and transported outside the object to be treated by a gas flow [93]. Usually, the DBD plasma uses atmospheric air as the working gas, while jet plasma uses different working gases such as helium, nitrogen, argon, oxygen, etc. Different types of plasma jets can be designed with various electrode configurations, working gases and applied electrical parameters [94].

4.1. Plasma Generation and Treatment

The two most commonly used NTP (DBD and Jet) configurations are shown in Figures 4 and 5, respectively. However, the three different approaches that can be used to treat samples with NTP are explained below:

Direct plasma treatment—It uses the target area as a counter electrode where there is a homogenous generation of plasma with high concentrations of plasma-generated species [95], e.g., DBD plasma.

Indirect plasma treatment—In this treatment technique, the plasma produced between two electrodes are later transported to the target area either by a carrier gas or diffusion [95], e.g., plasma needles, jets and torches.

Hybrid plasma treatment—In this approach, the plasma is produced in multiple micros and nano-discharges in a grounded wire mesh electrode, e.g., surface micro discharge (SMD).

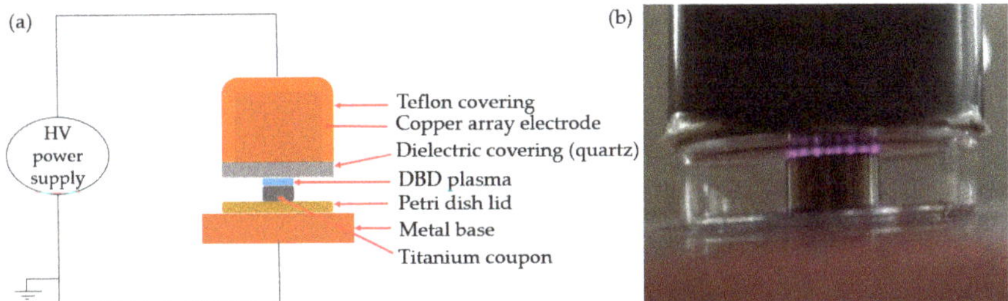

Figure 4. Schematic diagram and photograph of the regular DBD plasma setup. (**a**) demonstrates the schematic diagram of the regular DBD plasma and (**b**) shows the actual experimental setup of DBD plasma [70].

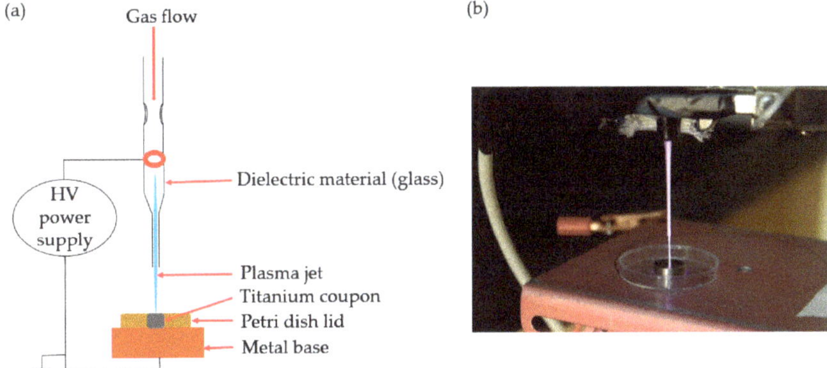

Figure 5. Schematic diagram and photograph of the jet plasma setup. (**a**) demonstrates the schematic diagram of the jet plasma and (**b**) shows the actual experimental setup of jet plasma [69,96].

4.2. Active Plasma Agents

The main active agents present in the plasma are radicals, charged particles, reactive oxygen species (ROS-O, O_2^*, O_3, OH), reactive nitrogen species (RNS-NO, NO_2), UV radiation, and electrical field. These active agents in combination are believed to be responsible for the antimicrobial efficacy. The reactive species (ROS and RNS) possess strong oxidative effects on the outer cellular structure [74]. ROS in cellular level leads to lipid peroxidation, DNA damage, protein modulation and programmed cell death in microorganisms [97,98]. Some of the molecular marker that is involved in plasma treatment are listed below in Table 3. UV radiation has fewer effects on bacteria as NTP at atmospheric pressure is a poor source of UV [97]. Similarly, charged particles play a vital role in rupturing the bacterial outer cell membrane. The electrostatic force created by the charge accumulation on the outer cell membrane overcomes its tensile strength and ruptures it [74].

Table 3. Molecular markers involved in plasma treatment.

Molecular Marker Involved in Plasma Treatment	Significance
8-hydroxydeoxyguanosine (8-OHdG) and Y-H2AX	Ubiquitous marker of oxidative stress and a by-product of oxidative DNA damage [99,100]
3-nitrotyrosine	Protein oxidative damage marker causes chemical fragmentation, inactivation, and proteolytic degradation [100]
Proteinase K	Plasma exposure reduces the catalytic activity of proteinase K by damaging the protein [101]
Malondialdehyde (MDA) and 4-hydroxynonenal (4-HNE)	A marker for oxidative stress that measures lipid peroxidation. It damages DNA and proteins through the formation of covalent adducts [99,102]
Polyunsaturated fatty acids (PUFA)	Causes lipid peroxidation of bacterial cell membrane by extracting H atom from PUFA by plasma Reactive oxygen species (ROS) [98]
Intracellular ATP	Poly(ADP-ribose) polymerase-1 (PARP-1) results in decreased ATP level which signifies cell surface damage caused by leaking cellular proteins/nucleic acids [103,104]

5. Factors Influencing the Antimicrobial Efficacy of NTP

Sterilization of biofilm via NTP is promising. However, it is affected by various factors, and careful consideration should be given while designing the plasma system for this purpose. Some of the factors that affect the efficacy of NTP are discussed in the following sections.

5.1. Plasma Treatment Time and the Distance between the Plasma Source and the Sample

NTP is regarded as dose-dependent, i.e., its efficacy depends on the plasma treatment time and the distance between the plasma nozzle and the sample as shown by one of the study [52]. This study found an increase in survival rate of *S. aureus* biofilm upon increasing the distance between the plasma source and the sample. More biofilm was killed at 8 mm distance in compared to the other distances used, such as 9, 10, and 11 mm. One of the potential reasons could be the reactive species that can reach the sample depend upon the distance, at fixed flow rate of working gas [105]. The antimicrobial efficacy of plasma increases with plasma treatment time and remains constant after a certain exposure period as revealed by several studies [3,59,106]. Different doses of photon and reactive species can be directed at the target sample by varying the distance between the plasma nozzle and the sample [107].

5.2. Frequency and Electrical Input Power (Voltage)

Electrical input power is another parameter which can be optimized to increase the antimicrobial efficacy of NTP [108,109]. The plasma power mainly depends on the distance between the plasma nozzle and the sample [78]. The amount of photons generated per second in a defined volume increases with increasing power [108]. One of the studies demonstrated greater reduction of *E. coli* and *L. monocytogenes* viability when using a higher voltage of 70 kV$_{\text{RMS}}$ after plasma treatment [110]. They found significant effects on cell integrity when higher voltage was used with a shorter treatment time of 5 sec. Another study using DBD plasma of 20 kV and 25 kV resulted 2.43 and 4.12 log reduction in bacterial cells while those with 16 kV showed 1.0 log reduction. The reason for this is the production of higher input energy density with higher voltage [98]. Frequency also plays a major role in increasing the plasma efficacy. One research group showed the complete eradication of *P. aeruginosa* biofilm after increasing the frequency from 20 kHz to 40 kHz. This higher frequency results in a higher density of the plasma reactive species delivered to the target by generating more plasma pulses and effective plasma on-time [3].

5.3. Role of the Gas or the Gas Mixture with its Flow Rate

As mentioned earlier, different gases and mixtures have been used in plasma jets to increase antimicrobial efficacy. Argon requires lower ionization energy than helium and it is cheaper, whereas helium possesses better thermal conductivity thereby preventing thermal instabilities [107]. Moreover, the noble gases increase the antimicrobial properties and plasma stability, whereas the addition of oxygen aids in producing chemically active species [111–113]. One study demonstrated more inactivation of spores of *Bacillus* genus when He was combined with 3% O_2 in comparison to 100% He [114]. This is due to the generation of oxygen species such as singlet O and O_3 in a He/O_2 plasma environment, as mentioned by the study [114]. These oxygen species play a significant role in the sterilization process because these species have strong oxidative effects on the outer membrane of the bacterial cell [39,54,115]. The addition of oxygen in the working gas contributes to the deactivation efficacy of the plasma jet by the combined action of plasma-induced endogenous ROS and the plasma generated ROS as reveal by the study [116]. This combined effect damages the bacterial membrane leading to biofilm deactivation [116]. Nitrogen has also been added to the noble gases to increase the formation of reactive nitrogen species [107]. Up to 3% of oxygen maintains the discharge stability, which produces reactive metastables that are long-lived and capable of traveling tens of centimeters at the nozzle exit velocity. This occurs even though the usual lifetime of atomic oxygen is of the order of 1 ms [117]. However, if more oxygen is added to the helium, the discharge becomes unstable due to the quenching effect of the oxygen gas, and this would decrease the plasma efficacy as a result of a decrease in plasma density [111].

Many researchers have chosen argon as the working gas because of its inertness and have succeeded in achieving planktonic and biofilm sterilization [51,118–120]. One of the study observed the sterilization effects of microwave-induced argon plasma on *E. coli* and MRSA and suggested the generation of free radicals, and UV light, as well as the etching process, are responsible for the sterilization efficacy [118]. Coupled to effect of the choice of gas, its flow rate determines the velocity at which the active species are delivered to the target [121]. The study by Nishime et al., uses a He plasma with a flow rate of 2 and 4 SLM and found that the shape of the inhibition zone and homogeneity were compromised at higher flows with no significant difference in the size of inhibition zones [121]. With higher gas flow rates, flow dynamics such as turbulent mixing and buoyancy effects play major roles in the formation and distribution of active species [122].

6. Biomedical Applications of NTP on Biofilm

A wide variety of research has been published on the use of NTP in biofilm treatment [3,53,61,62,120,123] with the help of different plasma systems such as corona discharge, microwave discharges, plasma jet, gliding arc, and dielectric barrier discharge within the past few years. The literature published in 2015 by Xu et al. demonstrated complete inactivation of *Staphylococcus aureus* biofilms that was grown on borosilicate slices placed in the 24-well plates for 12 h [49]. They used atmospheric pressure plasma jet (APPJ) with helium as a working gas at 6.7 standard liters per minute (SLM) flow rate. Within 10 min of plasma treatment, the reduction in biofilm cells was found to be more than 99.9% in comparison to the untreated biofilm samples. The effect of plasma treatment on biofilm was observed using a confocal microscope, exhibiting many dead bacteria on the biofilm upper layer in comparison to the bottom layer. Further, intra-bacterial ROS in the biofilm was detected by ROS monitoring probe 2, 7-dichlorodihydrofluorescein diacetate (DCFH-DA). Their findings also provide insight into the mechanism of biofilm inactivation by plasma reactive species and plasma-induced intracellular ROS.

Matthes et al. in 2013 studied the antimicrobial efficacy of two surface barrier discharges (SBD) known as SBD-A (structured electrode planar SBD) and SBD-B (a wire electrode SBD), with air plasma in *Pseudomonas aeruginosa* and *Staphylococcus epidermidis* biofilms. The biofilm was grown on polycarbonate discs placed into microplate wells for 48 h and treated with plasma from 30–600 s. They achieved a colony reduction factor (CRF) of 7.1 $\log_{10}$ and 3.81 $\log_{10}$ for *P. aeruginosa* by SBD-A and

SBD-B plasma. Whereas for *Staphylococcus epidermidis*, CRF of 3.38 $\log_{10}$ and 2.69 $\log_{10}$ were found out by SBD-A and SBD-B plasma at 600 s of plasma exposure. The study used a positive control as CHX that is used as a treatment for dental biofilm and found around 1 log reduction in biofilm cells. Furthermore, in a similar study, the cytotoxicity of the plasma system was tested on a mouse fibroblast cell line concluding that the average viability of the cells did not decrease below 50% until 150 s of plasma exposure time [53].

Non-thermal plasma has also been used widely in dentistry to remove dental biofilms. A 2014 study used positive corona (PC) and negative corona (NC) discharges on biofilm contaminated teeth surfaces and tested the effect of water electro-spraying for decontamination [60]. Both discharges were able to reduce the biofilm cell concentration by 1–1.3 orders of magnitude and 2.73 logs in 5 and 10 min of plasma treatment time. In addition, 3.16 orders of biofilm cell reduction were achieved after water electro-spraying through the plasma. The impact of those discharges on tooth surfaces was also studied by FTIR and SEM, but no significant changes were observed. In 2010 and 2011 Koban et al. used three plasma devices named as atmospheric pressure plasma jet, a hollow dielectric barrier discharge electrode (HDBD) and a volume dielectric barrier discharge (VDBD) against dental biofilms. Their results showed log reduction of 5.38 and 5.67 for *Streptococcus mutans* (*S. mutans*) and saliva biofilm when compared with the CHX as a control which shows a reduction of 3.36 and 1.50 for *S. mutans* and saliva biofilm [31]. They also achieved 5 log reduction of *Candida albicans* biofilm cells and suggested the plasma to be more effective than CHX in the treatment of single and multispecies dental biofilms.

In comparison to the plasma treated biofilm that has a 5 log reduction, the chemical antiseptics such as CHX or NaOCl had a reduction factor of 1.5, suggesting that this plasma can be used as an alternative to chemical antiseptics for dental practice [54]. This study concluded the plasma could be used as an alternative to chemical antiseptics for dental applications. One of our study [70] showed 2.43 log reduction of biofilm when treated with CHX in compare to more than 3 log reduction after plasma treatment. We also found some disruption of biofilm cells when treated with CHX, however the biofilm destruction caused by plasma was more severe. A 2010 study used glass coverslips for growing biofilm and achieved complete inactivation (7 log reduction) of *Neisseria gonorrhoeae* biofilm after treating with jet plasma for 20 min [42]. A mixture of helium and oxygen were used as a working gas to generate plasma. In the similar study, the effect of plasma on biofilm was visualized by transmission electron microscopy (TEM). The images show disruption and damage to the cell wall and dispersed nucleoid region.

In 2014, Vandervoort et al. established complete inactivation of *Pseudomonas aeruginosa* biofilm after 30 min of plasma jet treatment. The biofilm in this study was grown for 24 h on borosilicate glass coupons under continuous culture system [56]. Their atomic force microscopy (AFM) results show significant loss of the biofilm structure when treated with plasma for a longer time. They also concluded that changes in biofilm structure that leads to the loss of culturability and viability are related to a decrease of the biofilm matrix adhesiveness. Similarly, Zelaya et al. claimed 100% inactivation of the *P. aeruginosa* biofilm cells after 5 min of applying plasma jet. They used batch culture to grow 1, 3, and 7 day old biofilm on borosilicate glass coupons [62]. They also showed a decrease in adhesiveness to borosilicate and biofilm thickness after plasma treatment by AFM.

Another study by Alkawareek et al. in 2012 achieved complete eradication and more than 4 log (99.99%) reduction in the number of biofilm cells when treated with 40 and 20 kHz plasma jets. Their confocal microscopy results indicate this from the biofilm thickness after 240 s of plasma treatment suggesting the penetration of reactive species into the biofilm inner layer. This study demonstrated the potential to completely remove biofilms formed on inanimate surfaces employed in the manufacture of indwelling medical devices [3]. Similarly, another study in 2012 showed 6 log reduction of *Candida albicans* biofilms after treating with surface micro-discharge (SMD) plasma for 8 min. This contact-free application of plasma to kill biofilm cells could be beneficial for the eradication of nosocomial infections in hospital settings [55].

Liu et al. in 2017 evaluated the bactericidal effects of non-thermal argon/oxygen plasma on *S. mutans* and/or *S. sanguinis* biofilms [124]. The 7 days old biofilm grown on 48-well plate showed 99% of bacterial cells reduction after 2 min plasma treatment. Their study further suggested the virulence properties of *A. oris* by altering its hydrophobicity and capability to co-aggregate with *S. sanguinis*. A study by Bhatt et al. in 2018 used a novel argon plasma-activated gas (PAG) for the treatment of different biofilm species such as *S. aureus*, *P. aeruginosa*, and *E. coli*. These biofilms were grown on polytetrafluoroethylene channel segments similar to the GI endoscopic channels for 48 h and treated with PAG for 9 min. Their results demonstrated effective killing of biofilm (8 log reduction) that have a potential alternative to the high-level disinfectants (HLDs) and/or ethylene oxide in the endoscope reprocessing procedure [125].

A study by Patenall et al. in 2018 investigated the ability to disrupt and limit biofilms growth of *P. aeruginosa* by helium cold atmospheric pressure plasma (CAP) jet. 4–5 log reduction in viable bacterial cells were found when 8 h grown biofilms were treated with plasma, whereas only 2 log reduction were achieved after treating biofilms grown for 12 h. The plasma treatment time was 5 min. This study concludes that using CAP in a time-dependent manner is important for reducing the formation of biofilms [126]. Another interesting study by Lu et al. in 2018 used helium porous plasma jet on *P. aeruginosa* biofilms grown on the surface of the glass vial for 24 h. The results showed 4.5 $\log_{10}$ reduction of bacterial cells after 5 min of plasma treatment. This could provide an alternative approach for inner surface sterilization and decontamination in the medical device, food, and pharmaceutical industries [127].

Thus, the most commonly used plasma configurations are jet and DBD plasma. In DBD configuration, the plasma ignites from an electrode array, therefore, covering larger area and can treat larger surfaces for planktonic as well as bacteria in biofilms. This plasma is more suitable in clinical settings to treat the biofilm contaminated medical devices and implants. However, jet plasma has been widely used when compared to the DBD as previously mentioned. Jet configuration is relatively smaller and is suitable for treating bacteria in nooks and crannies of a surface, for example, in dentistry. This is possible since a plasma source with jet configuration can be customized in a desired shape and size depending upon the application. It is also possible to use different gases for plasma generation in this configuration leading to the generation of more ROS and RNS. When comparing the inactivation yield, jet configuration has demonstrated better efficacy than the DBD configuration. The jet plasma can yield complete sterilization of biofilm [56], however, the maximum inactivation yield by DBD plasma was only 7.1 log reduction [53]. In both cited studies, the bacterial strain used was *Pseudomonas aeruginosa*. On the other hand, it has been also found that the performance of NTP depends upon the parameters used. For example, the study [42] used 10 kV and 10 kHz to generate jet plasma and yielded 7 log reduction, whereas, another study [3] yielded only 4 log reduction when treating with plasma generated by 6 kV and 20 kHz source. A similar difference was observed when comparing the results of other two studies on DBD plasma. One of the studies [53] demonstrated 3–7 log reduction when treated with DBD plasma from 8–13 kV and 20–30 kHz, however, there was only 5 log reduction when treated with 8–10 kV and 37–40 kHz plasma source [54]. Therefore, it can be concluded that the antibacterial efficacy of plasma depends on several parameters like plasma generation, biological, and environmental parameters. Some of the plasma parameters are exposure time, voltage, frequency, distance between plasma source and the sample, working gas type, and the flow rate. The biological parameters that might affect the efficacy could be the bacterial species of interest and gram characteristics, initial inoculum, growth stages, biofilm growth period and mode, and the thickness of biofilm. Similarly, the environmental parameter affecting the inactivation yield by plasma could be the matrix composition, its relative humidity (RH), and the acidity [98].

7. Other Medical Applications

The advent of NTP has opened multiple opportunities for its use in several areas. Apart from biofilm decontamination and sterilization, it is being widely investigated in cancer treatment, blood coagulation, wound healing, tissue regeneration, and dentistry. NTP causes selective death (apoptosis) of cancer cells as these cells are more sensitive to plasma treatment than non-malignant ones [93]. This anti-cancer effect has also been demonstrated in vivo [93].

Dentistry is another field where plasma has been used to treat dental implants, intraoral disease, and cleaning or disinfecting dental cavity tissue or tooth root canal [86]. Blood coagulation has also been possible using plasma without any thermal effects and bacterial contamination [72,128]. Some of the findings using skin models suggest that using NTP can contribute to decontaminating acute and chronic wounds and accelerate healing [129]. These advances in plasma surgery wound healing, and tissue regeneration is because of the development of a plasma device called "Plazon" [39] that has been in medical use for patients for nine years [130].

8. Conclusions

The extensive studies of recent years clearly show the promising nature of NTP technology in eradicating biofilms in the medical field. However, there is a long unmet need for the development of a robust platform that can utilize the advantages of this technology. Therefore, the development of feasible NTP devices is needed if this technology is to make a paradigm shift in the world of decontamination/sterilization rather than being currently limited to in vitro and a few in vivo studies.

Author Contributions: Conceptualization and methodology, T.T.G. and H.A.; critical analysis, writing, and original draft preparation, T.T.G.; review and editing, T.T.G. and H.A.

Funding: This research received no external funding.

Acknowledgments: We would like to thank Arunan Nadarajah for the editorial assistance with this manuscript.

Conflicts of Interest: The authors declare no conflicts of interest.

References

1. Smith, A.W. Biofilms and antibiotic therapy: Is there a role for combating bacterial resistance by the use of novel drug delivery systems? *Adv. Drug Deliv. Rev.* **2005**, *57*, 1539–1550. [CrossRef] [PubMed]
2. Puligundla, P.; Mok, C. Potential applications of nonthermal plasmas against biofilm-associated micro-organisms in vitro. *J. Appl. Microbiol.* **2017**, *122*, 1134–1148. [CrossRef] [PubMed]
3. Alkawareek, M.Y.; Algwari, Q.T.; Laverty, G.; Gorman, S.P.; Graham, W.G.; O'Connell, D.; Gilmore, B.F. Eradication of Pseudomonas aeruginosa biofilms by atmospheric pressure non-thermal plasma. *PLoS ONE* **2012**, *7*, e44289. [CrossRef] [PubMed]
4. Høiby, N.; Bjarnsholt, T.; Givskov, M.; Molin, S., Ciofu, O. Antibiotic resistance of bacterial biofilms. *Int. J. Antimicrob Agents* **2010**, *35*, 322–332. [CrossRef] [PubMed]
5. Wu, H.; Moser, C.; Wang, H.-Z.; Høiby, N.; Song, Z.-J. Strategies for combating bacterial biofilm infections. *Int. J. Oral Sci.* **2015**, *7*, 1. [CrossRef] [PubMed]
6. Cornell, R.S.; Hood, C.R. What You Should Know About Biofilm and Implants. *Podiatry Today* **2015**, *28*, 38–46. Available online: https://www.podiatrytoday.com/what-you-should-know-about-biofilm-and-implants (accessed on 15 August 2019).
7. Wolcott, R.; Dowd, S. The role of biofilms: Are we hitting the right target? *Plast. Reconstr. Surg.* **2011**, *127*, 28S–35S. [CrossRef]
8. Abreu, A.C.; Tavares, R.R.; Borges, A.; Mergulhão, F.; Simões, M. Current and emergent strategies for disinfection of hospital environments. *J. Antimicrob. Chemother.* **2013**, *68*, 2718–2732. [CrossRef]
9. Khardori, N.; Yassien, M. Biofilms in device-related infections. *J. Ind. Microbiol. Biotechnol.* **1995**, *15*, 141–147. [CrossRef]
10. Donlan, R.M. Biofilms and device-associated infections. *Emerg. Infect. Dis.* **2001**, *7*, 277. [CrossRef]
11. Murakami, M.; Nishi, Y.; Seto, K.; Kamashita, Y.; Nagaoka, E. Dry mouth and denture plaque microflora in complete denture and palatal obturator prosthesis wearers. *Gerodontology* **2015**, *32*, 188–194. [CrossRef] [PubMed]
12. Song, Z.; Borgwardt, L.; Høiby, N.; Wu, H.; Sørensen, T.S.; Borgwardt, A. Prosthesis infections after orthopedic joint replacement: The possible role of bacterial biofilms. *Orthop. Rev.* **2013**, *5*, 14. [CrossRef] [PubMed]

13. Tran, P.L.; Lowry, N.; Campbell, T.; Reid, T.W.; Webster, D.R.; Tobin, E.; Aslani, A.; Mosley, T.; Dertien, J.; Colmer-Hamood, J.A. An organoselenium compound inhibits Staphylococcus aureus biofilms on hemodialysis catheters in vivo. *Antimicrob. Agents Chemother.* **2012**, *56*, 972–978. [CrossRef] [PubMed]

14. Santos, A.P.A.; Watanabe, E.; Andrade, D. de Biofilm on artificial pacemaker: Fiction or reality? *Arq. Bras. Cardiol.* **2011**, *97*, e113–e120. [CrossRef] [PubMed]

15. O'Connor, N.; Cahill, O.; Daniels, S.; Galvin, S.; Humphreys, H. Cold atmospheric pressure plasma and decontamination. Can it contribute to preventing hospital-acquired infections? *J. Hosp. Infect.* **2014**, *88*, 59–65. [CrossRef] [PubMed]

16. Khan, M.S.A.; Ahmad, I.; Sajid, M.; Cameotra, S.S. Current and emergent control strategies for medical biofilms. In *Antibiofilm Agents*; Springer: Berlin/Heidelberg, Germany, 2014; pp. 117–159.

17. Rutala, W.A.; Weber, D.J. Guideline for Disinfection and Sterilization in Healthcare Facilities. 2008. Available online: http://www.cdc.gov/hicpac/pdf/guidelines/Disinfection_Nov_2008.pdf (accessed on 15 August 2019).

18. Weinstein, R.A.; Hota, B. Contamination, disinfection, and cross-colonization: Are hospital surfaces reservoirs for nosocomial infection? *Clin. Infect. Dis.* **2004**, *39*, 1182–1189. [CrossRef] [PubMed]

19. Cahill, O.J.; Claro, T.; O'Connor, N.; Cafolla, A.A.; Stevens, N.T.; Daniels, S.; Humphreys, H. Cold air plasma to decontaminate inanimate surfaces of the hospital environment. *Appl. Environ. Microbiol.* **2014**, *80*, 2004–2010. [CrossRef]

20. Lindsay, D.; Von Holy, A. Bacterial biofilms within the clinical setting: What healthcare professionals should know. *J. Hosp. Infect.* **2006**, *64*, 313–325. [CrossRef]

21. Bose, S.; Ghosh, A.K. Fungal Biofilm & Medical Device Associated Infection: It's Formation, Diagnosis & Future Trends: A Review. *Ann. Pathol. Lab. Med.* **2016**, *3*, R1–R9.

22. Brelles-Mariño, G. Challenges in biofilm inactivation: The use of cold plasma as a new approach. *J. Bioprocess. Biotech.* **2012**, *2*, e107.

23. Donlan, R.M. Biofilm formation: A clinically relevant microbiological process. *Clin. Infect. Dis.* **2001**, *33*, 1387–1392. [CrossRef] [PubMed]

24. McConoughey, S.J.; Howlin, R.; Granger, J.F.; Manring, M.M.; Calhoun, J.H.; Shirtliff, M.; Kathju, S.; Stoodley, P. Biofilms in periprosthetic orthopedic infections. *Future Microbiol.* **2014**, *9*, 987–1007. [CrossRef] [PubMed]

25. Hall-Stoodley, L.; Stoodley, P. Developmental regulation of microbial biofilms. *Curr. Opin. Biotechnol.* **2002**, *13*, 228–233. [CrossRef]

26. Izano, E.A.; Amarante, M.A.; Kher, W.B.; Kaplan, J.B. Differential roles of poly-N-acetylglucosamine surface polysaccharide and extracellular DNA in Staphylococcus aureus and Staphylococcus epidermidis biofilms. *Appl. Environ. Microbiol.* **2008**, *74*, 470–476. [CrossRef] [PubMed]

27. Høiby, N.; Ciofu, O.; Johansen, H.K.; Song, Z.; Moser, C.; Jensen, P.Ø.; Molin, S.; Givskov, M.; Tolker-Nielsen, T.; Bjarnsholt, T. The clinical impact of bacterial biofilms. *Int. J. Oral Sci.* **2011**, *3*, 55. [CrossRef]

28. Mah, T.-F.C.; O'Toole, G.A. Mechanisms of biofilm resistance to antimicrobial agents. *Trends Microbiol.* **2001**, *9*, 34–39. [CrossRef]

29. Stewart, P.S.; Costerton, J.W. Antibiotic resistance of bacteria in biofilms. *Lancet* **2001**, *358*, 135–138. [CrossRef]

30. Rupf, S.; Lehmann, A.; Hannig, M.; Schäfer, B.; Schubert, A.; Feldmann, U.; Schindler, A. Killing of adherent oral microbes by a non-thermal atmospheric plasma jet. *J. Med. Microbiol.* **2010**, *59*, 206–212. [CrossRef]

31. Koban, I.; Holtfreter, B.; Hübner, N.O.; Matthes, R.; Sietmann, R.; Kindel, E.; Weltmann, K.D.; Welk, A.; Kramer, A.; Kocher, T. Antimicrobial efficacy of non-thermal plasma in comparison to chlorhexidine against dental biofilms on titanium discs in vitro—Proof of principle experiment. *J. Clin. Periodontol.* **2011**, *38*, 956–965. [CrossRef]

32. Gupta, T.T.; Ayan, H. Non-Thermal Plasma in Conjunction with Chlorhexidine (Chx) Digluconate Sterilize The Biofilm Contaminated Titanium Surface. In Proceedings of the 2017 IEEE International Conference on Plasma Science (ICOPS), Atlantic City, NJ, USA, 21–25 May 2017; p. 1.

33. Gupta, T.T. Characterization and Optimization of Non-Thermal Plasma for Biofilm Sterilization. Ph.D. Thesis, University of Toledo, Toledo, OH, USA, 2018.

34. Jha, N.; Ryu, J.J.; Choi, E.H.; Kaushik, N.K. Generation and role of reactive oxygen and nitrogen species induced by plasma, lasers, chemical agents, and other systems in dentistry. *Oxid. Med. Cell. Longev.* **2017**, *2017*, 7542540. [CrossRef]

73. Kong, M.G.; Kroesen, G.; Morfill, G.; Nosenko, T.; Shimizu, T.; Van Dijk, J.; Zimmermann, J.L. Plasma medicine: An introductory review. *New J. Phys.* **2009**, *11*, 115012. [CrossRef]

74. Laroussi, M. Low temperature plasma-based sterilization: Overview and state-of-the-art. *Plasma Process. Polym.* **2005**, *2*, 391–400. [CrossRef]

75. Isbary, G.; Stolz, W.; Shimizu, T.; Monetti, R.; Bunk, W.; Schmidt, H.-U.; Morfill, G.E.; Klämpfl, T.G.; Steffes, B.; Thomas, H.M. Cold atmospheric argon plasma treatment may accelerate wound healing in chronic wounds: Results of an open retrospective randomized controlled study in vivo. *Clin. Plasma Med.* **2013**, *1*, 25–30. [CrossRef]

76. Ayan, H.; Fridman, G.; Gutsol, A.F.; Vasilets, V.N.; Fridman, A.; Friedman, G. Nanosecond-pulsed uniform dielectric-barrier discharge. *IEEE Trans. Plasma Sci.* **2008**, *36*, 504–508. [CrossRef]

77. Sanaei, N.; Ayan, H. Bactericidal efficacy of dielectric barrier discharge plasma on methicillin-resistant staphylococcus aureus and Escherichia coli in planktonic phase and colonies in vitro. *Plasma Med.* **2015**, *5*, 1–16. [CrossRef]

78. Stoffels, E.; Kieft, I.E.; Sladek, R.E.J.; Van den Bedem, L.J.M.; Van der Laan, E.P.; Steinbuch, M. Plasma needle for in vivo medical treatment: Recent developments and perspectives. *Plasma Sources Sci. Technol.* **2006**, *15*, S169. [CrossRef]

79. Sladek, R.E.J.; Stoffels, E.; Walraven, R.; Tielbeek, P.J.A.; Koolhoven, R.A. Plasma treatment of dental cavities: A feasibility study. *IEEE Trans. Plasma Sci.* **2004**, *32*, 1540–1543. [CrossRef]

80. Fridman, G.; Shereshevsky, A.; Jost, M.M.; Brooks, A.D.; Fridman, A.; Gutsol, A.; Vasilets, V.; Friedman, G. Floating electrode dielectric barrier discharge plasma in air promoting apoptotic behavior in melanoma skin cancer cell lines. *Plasma Chem. Plasma Process.* **2007**, *27*, 163–176. [CrossRef]

81. Karki, S.B.; Gupta, T.T.; Yildirim-Ayan, E.; Eisenmann, K.M.; Ayan, H. Investigation of non-thermal plasma effects on lung cancer cells within 3D collagen matrices. *J. Phys. D Appl. Phys.* **2017**, *50*, 315401. [CrossRef]

82. Karki, S.B.; Yildirim-Ayan, E.; Eisenmann, K.M.; Ayan, H. Miniature Dielectric Barrier Discharge Nonthermal Plasma Induces Apoptosis in Lung Cancer Cells and Inhibits Cell Migration. *BioMed Res. Int.* **2017**, *2017*, 8058307. [CrossRef]

83. Kamath, U.; Rajasekhara, S.; Nazar, N.; Sathar, A. Cold Atmospheric Plasma; Break Through in Dentistry—A Review. *Paripex-Indian J. Res.* **2016**, *5*.

84. Singh, S.; Chandra, R.; Tripathi, S.; Rahman, H.; Tripathi, P.; Jain, A.; Gupta, P. The bright future of dentistry with cold plasma-review. *J. Dent. Med. Sci.* **2014**, *13*, 6–13. [CrossRef]

85. Cheruthazhekatt, S.; Černák, M.; Slavíček, P.; Havel, J. Gas plasmas and plasma modified materials in medicine. *J. Appl. Biomed.* **2010**, *8*, 55–66. [CrossRef]

86. Scholtz, V.; Pazlarova, J.; Souskova, H.; Khun, J.; Julak, J. Nonthermal plasma—A tool for decontamination and disinfection. *Biotechnol. Adv.* **2015**, *33*, 1108–1119. [CrossRef]

87. Von Woedtke, T.; Reuter, S.; Masur, K.; Weltmann, K.-D. Plasmas for medicine. *Phys. Rep.* **2013**, *530*, 291–320. [CrossRef]

88. Kogelschatz, U. Dielectric-barrier discharges: Their history, discharge physics, and industrial applications. *Plasma Chem. Plasma Process.* **2003**, *23*, 1–46. [CrossRef]

89. Ehlbeck, J.; Schnabel, U.; Polak, M.; Winter, J.; Von Woedtke, T.; Brandenburg, R.; Von dem Hagen, T.; Weltmann, K.D. Low temperature atmospheric pressure plasma sources for microbial decontamination. *J. Phys. D Appl. Phys.* **2010**, *44*, 13002. [CrossRef]

90. Laimer, J.; Störi, H. Recent Advances in the Research on Non-Equilibrium Atmospheric Pressure Plasma Jets. *Plasma Process. Polym.* **2007**, *4*, 266–274. [CrossRef]

91. Kogelschatz, U. Atmospheric-pressure plasma technology. *Plasma Phys. Control. Fusion* **2004**, *46*, B63. [CrossRef]

92. Neuber, J.U. Non-Thermal Atmospheric-Pressure Plasma for Sterilization of Surfaces and Biofilms. Master's Thesis, Old Dominion University, Norfolk, VA, USA, 2016.

93. Weltmann, K.D.; von Woedtke, T. Plasma medicine—Current state of research and medical application. *Plasma Phys. Control. Fusion* **2016**, *59*, 14031. [CrossRef]

94. Winter, J.; Brandenburg, R.; Weltmann, K.D. Atmospheric pressure plasma jets: An overview of devices and new directions. *Plasma Sources Sci. Technol.* **2015**, *24*, 64001. [CrossRef]

95. Isbary, G.; Shimizu, T.; Li, Y.-F.; Stolz, W.; Thomas, H.M.; Morfill, G.E.; Zimmermann, J.L. Cold atmospheric plasma devices for medical issues. *Expert Rev. Med. Devices* **2013**, *10*, 367–377. [CrossRef]

96. Gupta, T.; Karki, S.; Fournier, R.; Ayan, H. Mathematical Modelling of the Effects of Plasma Treatment on the Diffusivity of Biofilm. *Appl. Sci.* **2018**, *8*, 1729. [CrossRef]
97. Kieft, I.E. *Plasma Needle: Exploring Biomedical Applications of Non-Thermal Plasmas*; Technische Universiteit Eindhoven: Eindhoven, The Netherlands, 2005; ISBN 9038627378.
98. Liao, X.; Liu, D.; Xiang, Q.; Ahn, J.; Chen, S.; Ye, X.; Ding, T. Inactivation mechanisms of non-thermal plasma on microbes: A review. *Food Control* **2017**, *75*, 83–91. [CrossRef]
99. Joshi, S.G.; Cooper, M.; Yost, A.; Paff, M.; Ercan, U.K.; Fridman, G.; Friedman, G.; Fridman, A.; Brooks, A.D. Nonthermal dielectric-barrier discharge plasma-induced inactivation involves oxidative DNA damage and membrane lipid peroxidation in Escherichia coli. *Antimicrob. Agents Chemother.* **2011**, *55*, 1053–1062. [CrossRef] [PubMed]
100. Kalghatgi, S.U. *Mechanisms of Interaction of Non-Thermal Plasma with Living Cells*; Drexel University: Philadelphia, PA, USA, 2010.
101. Alkawareek, M.Y.; Gorman, S.P.; Graham, W.G.; Gilmore, B.F. Potential cellular targets and antibacterial efficacy of atmospheric pressure non-thermal plasma. *Int. J. Antimicrob. Agents* **2014**, *43*, 154–160. [CrossRef] [PubMed]
102. Flynn, P.B.; Gilmore, B.F. Understanding plasma biofilm interactions for controlling infection and virulence. *J. Phys. D Appl. Phys.* **2018**, *51*, 263001. [CrossRef]
103. Kvam, E.; Davis, B.; Mondello, F.; Garner, A.L. Nonthermal atmospheric plasma rapidly disinfects multidrug-resistant microbes by inducing cell surface damage. *Antimicrob. Agents Chemother.* **2012**, *56*, 2028–2036. [CrossRef]
104. Kaushik, N.K.; Ghimire, B.; Li, Y.; Adhikari, M.; Veerana, M.; Kaushik, N.; Jha, N.; Adhikari, B.; Lee, S.-J.; Masur, K. Biological and medical applications of plasma-activated media, water and solutions. *Biol. Chem.* **2018**, *400*, 39–62. [CrossRef] [PubMed]
105. Anzai, K.; Aoki, T.; Koshimizu, S.; Takaya, R.; Tsuchida, K.; Takajo, T. Formation of reactive oxygen species by irradiation of cold atmospheric pressure plasma jet to water depends on the irradiation distance. *J. Clin. Biochem. Nutr.* **2019**, *64*, 187–193. [CrossRef]
106. Barekzi, N.; Laroussi, M. Dose-dependent killing of leukemia cells by low-temperature plasma. *J. Phys. D Appl. Phys.* **2012**, *45*, 422002. [CrossRef]
107. Lackmann, J.-W.; Bandow, J.E. Inactivation of microbes and macromolecules by atmospheric-pressure plasma jets. *Appl. Microbiol. Biotechnol.* **2014**, *98*, 6205–6213. [CrossRef]
108. Halfmann, H.; Bibinov, N.; Wunderlich, J.; Awakowicz, P. A double inductively coupled plasma for sterilization of medical devices. *J. Phys. D Appl. Phys.* **2007**, *40*, 4145. [CrossRef]
109. Halfmann, H.; Denis, B.; Bibinov, N.; Wunderlich, J.; Awakowicz, P. Identification of the most efficient VUV/UV radiation for plasma based inactivation of Bacillus atrophaeus spores. *J. Phys. D Appl. Phys.* **2007**, *40*, 5907. [CrossRef]
110. Lu, H.; Patil, S.; Keener, K.M.; Cullen, P.J.; Bourke, P. Bacterial inactivation by high-voltage atmospheric cold plasma: Influence of process parameters and effects on cell leakage and DNA. *J. Appl. Microbiol.* **2014**, *116*, 784–794. [CrossRef] [PubMed]
111. Ren, Y.; Wang, C.; Qiu, Y. Aging of surface properties of ultra high modulus polyethylene fibers treated with He/O_2 atmospheric pressure plasma jet. *Surf. Coat. Technol.* **2008**, *202*, 2670–2676. [CrossRef]
112. Park, J.; Henins, I.; Herrmann, H.W.; Selwyn, G.S.; Jeong, J.Y.; Hicks, R.F.; Shim, D.; Chang, C.S. An atmospheric pressure plasma source. *Appl. Phys. Lett.* **2000**, *76*, 288–290. [CrossRef]
113. Kim, S.J.; Chung, T.H.; Bae, S.H.; Leem, S.H. Bacterial inactivation using atmospheric pressure single pin electrode microplasma jet with a ground ring. *Appl. Phys. Lett.* **2009**, *94*, 141502. [CrossRef]
114. Laroussi, M.; Leipold, F. Evaluation of the roles of reactive species, heat, and UV radiation in the inactivation of bacterial cells by air plasmas at atmospheric pressure. *Int. J. Mass Spectrom.* **2004**, *233*, 81–86. [CrossRef]
115. Thiyagarajan, M.; Sarani, A.; Nicula, C. Optical emission spectroscopic diagnostics of a non-thermal atmospheric pressure helium-oxygen plasma jet for biomedical applications. *J. Appl. Phys.* **2013**, *113*, 233302. [CrossRef]
116. Xu, Z.; Shen, J.; Cheng, C.; Hu, S.; Lan, Y.; Chu, P.K. In vitro antimicrobial effects and mechanism of atmospheric-pressure He/O_2 plasma jet on Staphylococcus aureus biofilm. *J. Phys. D Appl. Phys.* **2017**, *50*, 105201. [CrossRef]

117. Herrmann, H.W.; Henins, I.; Park, J.; Selwyn, G.S. Decontamination of chemical and biological warfare (CBW) agents using an atmospheric pressure plasma jet (APPJ). *Phys. Plasmas* **1999**, *6*, 2284–2289. [CrossRef]

118. Lee, K.-Y.; Park, B.J.; Lee, D.H.; Lee, I.-S.; Hyun, S.O.; Chung, K.-H.; Park, J.-C. Sterilization of Escherichia coli and MRSA using microwave-induced argon plasma at atmospheric pressure. *Surf. Coat. Technol.* **2005**, *193*, 35–38. [CrossRef]

119. Park, B.J.; Lee, D.H.; Park, J.-C.; Lee, I.-S.; Lee, K.-Y.; Hyun, S.O.; Chun, M.-S.; Chung, K.-H. Sterilization using a microwave-induced argon plasma system at atmospheric pressure. *Phys. Plasmas* **2003**, *10*, 4539–4544. [CrossRef]

120. Lee, M.H.; Park, B.J.; Jin, S.C.; Kim, D.; Han, I.; Kim, J.; Hyun, S.O.; Chung, K.-H.; Park, J.-C. Removal and sterilization of biofilms and planktonic bacteria by microwave-induced argon plasma at atmospheric pressure. *New J. Phys.* **2009**, *11*, 115022. [CrossRef]

121. Nishime, T.M.C.; Borges, A.C.; Koga-Ito, C.Y.; Machida, M.; Hein, L.R.O.; Kostov, K.G. Non-thermal atmospheric pressure plasma jet applied to inactivation of different microorganisms. *Surf. Coat. Technol.* **2017**, *312*, 19–24. [CrossRef]

122. Goree, J.; Liu, B.; Drake, D. Gas flow dependence for plasma-needle disinfection of S. mutans bacteria. *J. Phys. D Appl. Phys.* **2006**, *39*, 3479. [CrossRef]

123. Sladek, R.E.J.; Filoche, S.K.; Sissons, C.H.; Stoffels, E. Treatment of Streptococcus mutans biofilms with a nonthermal atmospheric plasma. *Lett. Appl. Microbiol.* **2007**, *45*, 318–323. [CrossRef] [PubMed]

124. Liu, T.; Wu, L.; Babu, J.P.; Hottel, T.L.; Garcia-Godoy, F.; Hong, L. Effects of atmospheric non-thermal argon/oxygen plasma on biofilm viability and hydrophobicity of oral bacteria. *Am. J. Dent.* **2017**, *30*, 52–56.

125. Bhatt, S.; Mehta, P.; Chen, C.; Schneider, C.L.; White, L.N.; Chen, H.-L.; Kong, M.G. Efficacy of low-temperature plasma-activated gas disinfection against biofilm on contaminated GI endoscope channels. *Gastrointest. Endosc.* **2019**, *89*, 105–114. [CrossRef]

126. Patenall, B.L.; Hathaway, H.; Sedgwick, A.C.; Thet, N.T.; Williams, G.T.; Young, A.E.; Allinson, S.L.; Short, R.D.; Jenkins, A.T.A. Limiting Pseudomonas aeruginosa Biofilm Formation Using Cold Atmospheric Pressure Plasma. *Plasma Med.* **2018**, *8*, 269–277. [CrossRef]

127. Lu, P.; Ziuzina, D.; Cullen, P.J.; Bourke, P. Inner surface biofilm inactivation by atmospheric pressure helium porous plasma jet. *Plasma Process. Polym.* **2018**, *15*, e1800055. [CrossRef]

128. Kalghatgi, S.U.; Fridman, G.; Cooper, M.; Nagaraj, G.; Peddinghaus, M.; Balasubramanian, M.; Vasilets, V.N.; Gutsol, A.F.; Fridman, A.; Friedman, G. Mechanism of blood coagulation by nonthermal atmospheric pressure dielectric barrier discharge plasma. *IEEE Trans. Plasma Sci.* **2007**, *35*, 1559–1566. [CrossRef]

129. Lloyd, G.; Friedman, G.; Jafri, S.; Schultz, G.; Fridman, A.; Harding, K. Gas plasma: Medical uses and developments in wound care. *Plasma Process. Polym.* **2010**, *7*, 194–211. [CrossRef]

130. Heinlin, J.; Morfill, G.; Landthaler, M.; Stolz, W.; Isbary, G.; Zimmermann, J.L.; Shimizu, T.; Karrer, S. Plasma medicine: Possible applications in dermatology. *JDDG* **2010**, *8*, 968–976. [CrossRef] [PubMed]

 *applied sciences*

Article

Dry Bio-Decontamination Process in Reduced-Pressure O$_2$ Plasma

Hongxia Liu *, Xinxin Feng, Xin Ma, Jinzhuo Xie and Chi He

Department of Environmental Science and Engineering, School of Energy and Power Engineering,
Xi'an Jiaotong University, Xi'an 710049, China; fengxin2016@stu.xjtu.edu.cn (X.F.);
andrew995522@stu.xjtu.edu.cn (X.M.); xjz9611@stu.xjtu.edu.cn (J.X.); chi_he@xjtu.edu.cn (C.H.)
* Correspondence: hxliu72@xjtu.edu.cn; Tel.: +86-298-266-8572

Received: 21 March 2019; Accepted: 7 May 2019; Published: 10 May 2019

Featured Application: The specific findings of this research will deep our understanding for the mechanisms of bio-decontamination from material surfaces by plasma technology.

Abstract: The main objective of this work was to fully understand the bio-decontamination process in a reduced-pressure oxygen plasma. Gram-negative *Escherichia coli* species was chosen as the target microorganism in this test. The comparison of decontamination efficacy between plasma total and UV radiation individually under various treatment parameters and tests of DNA agarose electrophoresis were made to evaluate the inactivation effect of UV radiation. The quantity of protein leakage and the concentration of malondialdehyde (MDA), which are markers of the end products of lipid peroxidation, in bacterial suspension after treatment were determined to estimate the contribution of both charged particles and free radicals for bacterial death. In addition, a scanning electronic microscope was used to visualize the plasma effect on microorganisms. The results showed that the essential action of the oxygen plasma on *Escherichia coli* is believed to be attributed to the fast and intense etching on cell membrane by electrons and ions. Attacks on polyunsaturation fatty acid (PUFA) in the cell membrane by oxygen free radicals and the destruction of the DNA in the cell by UV radiation are accessorial during an effective decontamination process.

Keywords: oxygen plasma; *Escherichia coli*; bio-decontamination

1. Introduction

The elimination of disease-causing agents from the surfaces of equipment, which can sometimes be challenging to fulfill without using toxic materials or high temperatures, is an absolutely necessary requirement in many fields. Historically, many different approaches have been used to inactivate pathogens. Two widely used inactivation methods, especially in the medical field, are autoclaving and exposure to gases such as ethylene oxide (EtO). Though effective, both methods suffer from drawbacks such as exposure to extremely high temperatures (>1000 °C) in the case of autoclaves, and toxic exposure in the case of EtO. Another concern with the latter is the long aeration process which, importantly, creates a serious threat for both personnel and the environment [1]. For these reasons, it is extremely important to develop a new bio-decontamination or disinfection technology that is safe and easy to apply. In recent years, one of the most serious current alternatives to gaseous bio-decontamination is the use of gas-discharge plasma as a sterilizing agent. Plasma-based bio-decontamination techniques do not suffer from the problems of traditional techniques. Adequate processes are efficient, do affect only slightly the bulk material, are environmentally sound, do not produce toxic by-products, and are fast and cost-effective. Therefore, this kind of method is regarded as a green technique and even as the most promising bio-decontamination technique [2–4].

During the plasma process, micro-organisms are exposed to reactive species, which are produced by applying electromagnetic fields to a gas or gas mixture. But so far, all the mechanisms which may be responsible for the treatment, such as the interaction of UV radiation with the DNA of cells, the removal of the material of cells by reactive species and so on [5–10], are only presumed on the basis of the apparent reactive results in plasma. There is no thorough understanding about the respective contribution of different reactive species for microbial inactivation. That is, the present conclusions about plasma bio-decontamination mechanisms, although various, have less persuasion. As we know, being a bio-decontamination technique, those with higher stability and security than other ones must be paid attention. It is absolutely necessary to first investigate the bio-decontamination mechanisms of gas-discharge plasma.

Generally speaking, during plasma bio-decontamination, the complex reactive species can be divided into three categories: charged species (including electrons and ions), free radicals, and UV radiation. The last, UV, is the only one can be separated from other species by a special fitting due to its penetrability. From current research [7,8,11–13], we can see that UV radiation reacts with DNA to injure the micro-organism; the charged particles remove the material (i.e., etching) of cells, resulting in protein leakage; and the activated free radicals initiate the lipid peroxidation by attacking polyunsaturation fatty acid (PUFA) in the cell membrane. In accordance with these reasons, in this paper, the comparison of germicidal efficacy under various treatment conditions between plasma and UV radiation and the test of DNA state are combined to determine the decontamination effect of UV radiation firstly; then the observation of bacterial ultrastructure and the quantity of protein leakage as well as the yield of malondialdehyde (MDA), being the marker of the end products of the lipid peroxidation, are measured to ascertain the efficiency of the other two kinds reactive species. Reduced-pressure oxygen plasma was applied to perform in this experiment and Gram-negative *Escherichia coli* (*E. coli*) 8099 was used as a test strain.

2. Materials and Methods

2.1. Materials

E. coli 8099 slant lawn incubated at 37 °C for 24 h was oscillated and eluted by phosphate buffer solution (PBS), then the eluent was diluted to form certain concentration suspensions of bacilli. This suspension was taken 0.01 mL to spread uniformly on 15 × 15 mm glass sheet and dried naturally at room temperature.

2.2. O_2 Plasma Treatment

Figure 1 shows the schematic diagram of experimental apparatus. An RF generator was the type of SY-500 W, used whose power was 13.56 MHz and whose output power could be adjusted in succession in order to match the SP-II matcher. The reaction chamber was a Pyrex glass tube (length 150 mm, diameter 45 mm), initiated inductively coupled discharge. In addition, a vacuum pump was used to insure the reduced pressure in reactor.

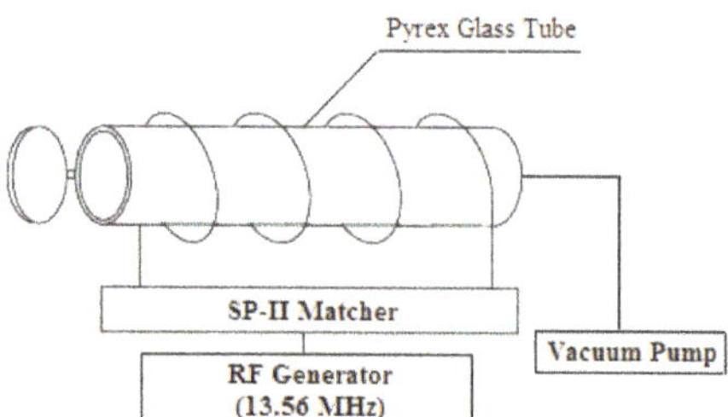

Figure 1. Schematic structure of plasma reactor for bio-decontamination from surface.

In the first test, some samples were directly exposed to plasma discharge zone to obtain the decontamination effects of *E. coli* by oxygen plasma in consideration of different operation parameters, such as gas flow—20–100 $cm^3 \cdot min^{-1}$, pressure—45 Pa, RF power—20–100 W and treatment time—20–120 s. In the second test, at the above-mentioned conditions, other samples were treated under the filter of lithium fluoride (LiF), which achieved the separation of UV and other active species, to determine the relative contribution of UV radiation to microbial inactivation. All factors of plasma generated reactive species (e.g., electrons, ion, free radicals and UV radiation) play roles in microbial inactivation in the former test, whereas in the latter test, microbial inactivation was performed with the action of UV radiation from plasma discharge with wavelength $\lambda \geq 120$ nm [14].

2.3. Viable Cell Counts and Decontamination Effect

Based on the living conditions of *E. coli*, the viable cell counts of teat samples was determined by cell culture methods, which was slightly modified from Hertwig et al. [15]. Simply, treated samples were withdrawn from the reactor immediately after the above-mentioned treatment. Then, eluting bacilli from control or treated samples with phosphate buffer solution (PBS) obtained the suspensions of bacilli, which were transferred to nutrient-containing Petri dish and incubated for 48 h at 37 °C prior to determining the bacilli number of colony forming units (CFU). The decontamination effect (E) of *E. coli* by oxygen plasma and UV radiation individual was determined by Equation (1) below [16–19],

$$E = \lg N_0 - \lg N_t \tag{1}$$

where N_0 and N_t represent the number of CFU of control and treated surviving bacilli, respectively.

2.4. Determination of Protein Leakage Quantity

The dye binding method was used as the detection of protein leakage in the cells after plasma treatment. The preparation of bacilli suspensions was described in Section 2.2. Then, the suspensions were placed in a beaker containing 5 mL of PBS, shaken for 10 min, and centrifuged at 3000 r/min for 15 min. We took 1.0 mL of the supernatant to mix with 5.0 mL of Coomassie Brilliant Blue dye, which was placed for 5 min at room temperature. The absorbency of the mixed solution was measured at the wavelength of 595 nm, meanwhile, obtained absorbance value was placed in the regression equation of standard curve of bovine serum albumin to obtain the quantity of protein leakage of cells after treatment [17,18,20].

2.5. Determination of Malondialdehyde (MDA) Production

Due to the condensation reaction of malondialdehyde (MDA) and thibabituric acid (TBA, obtained from Nanjing Jiancheng biological engineering institute) formed a red product with a maximum absorption peak at 532 nm, the determination of MDA production after treatment was obtained by colorimetry. The bacterial suspension was obtained by the same method as mentioned above. The MDA concentration was calculated by Equation (2) below [12,20,21],

$$N = \frac{A_m - A_{mo}}{A_s - A_{so}} \times 10^{-5} mol \cdot L^{-1} \times n \tag{2}$$

where N is the concentration of MDA ($mol \cdot L^{-1}$), A_m is determination tube absorbency, A_{mo} is determination blank tube absorbency, A_s is standard tube absorbency, A_{so} is s standard blank tube absorbency, n is the sample diluted multiple before test.

2.6. DNA Measurement

The DNA measurement of samples cells was extracted by distilling reagent of Genome DNA (TIANGEN Biological Technology Co., Ltd., Beijing, China), and measured by ECP3000 agarose electrophoresis apparatus (Six-one Instrument Co., Ltd., Beijing, China) [22]. Electrophoresis buffer was 0.5 × TBE, which pH was maintained at 7.5–7.8.

2.7. Scanning Electron Microscopy

The suspension of bacilli before and after inactivation was taken at 0.03 mL to spread uniformly on glass sheet to form folium with 1 cm diameter, and then dried naturally at room temperature. The ultrastructure of bacilli was observed using scanning electron microscopy (SEM) in a JEOL instrument (Model JSM–6700F, Tokyo, Japan) after coating the dry specimens with gold. The magnification of 10000× was used.

2.8. Statistical Analysis

In Sections 2.3–2.6 the determination of decontamination effect, protein leakage quantity, MDA production and DNA for samples were carried out in triplicate to ensure the repeatability. The data were expressed as the mean ± standard deviation (SD) in Figures 2, 4, and 6. The significant of differences among data were statistically analyzed by one-way ANOVA test using SPSS software (version 20.0), choosing $p < 0.05$ as statistically significant.

3. Results

3.1. Decontamination Effect of UV Radiation in Oxygen Plasma

The decontamination efficiency of oxygen plasma total and UV radiation individual (excluded the charge particles and free radicals generated by plasma discharge using the LiF) was investigated at the same operation parameters, respectively, from E as functions of plasma RF power, treatment time and gas flow rate. The samples were positioned at the center of the induction coil in the Pyrex glass tube of the reactor. Figure 2 shows the comparison of both the decontamination effects of oxygen plasma total and UV radiation individually. From Figure 2, for one thing, it could be clearly seen that the E of oxygen plasma total showed a strong dependence on the conditions mentioned above, and the effective bio-decontamination was carried out when oxygen plasma discharge under the conditions of 100 W, 60 s and 40 cm^3·min^{-1}, where the E was largest, up to 3.42.

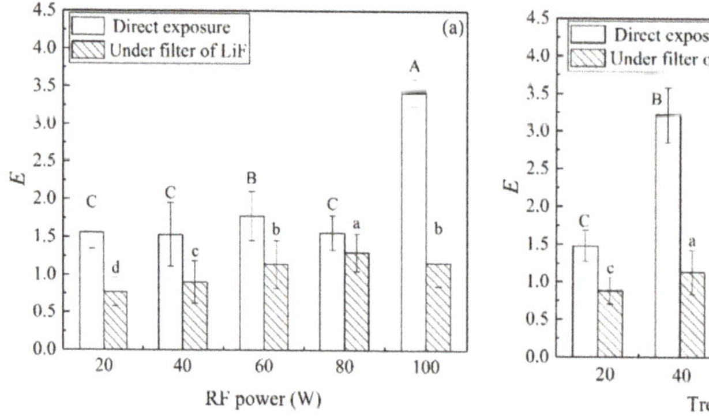

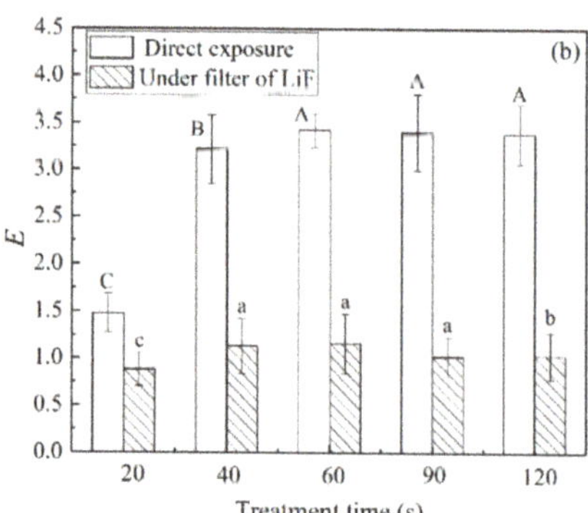

Figure 2. *Cont.*

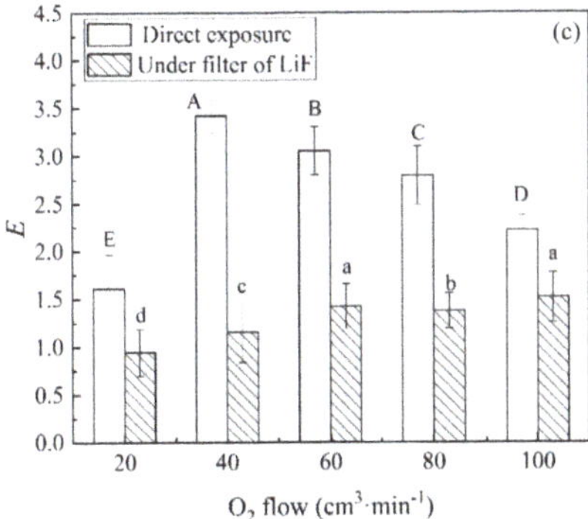

Figure 2. Comparison of the decontamination effect between oxygen plasma total and UV radiation individual under different treatment conditions (**a**) Time 60 s, O_2 flow 40 cm^3·min^{-1}, Pressure 45 Pa; (**b**) Power 100 W, O_2 flow 40 cm^3·min^{-1}, Pressure 45 Pa; (**c**) Power 100 W, Time 60 s, Pressure 45 Pa. Capital letters A–E are distinguished from a–d to represent statistical significance in direct exposure.

In addition, the decontamination rate of *E. coli* by oxygen plasma as a function of O_2 flow rate increased firstly ($p < 0.05$) and subsequently decreased ($p < 0.05$) with the increase of gas flow and was optimal at a flow rate of 40 cm^3·min^{-1}. This phenomenon was due to the inactivation effect of *E. coli* depending on the average energy, number, and residence time of reactive species in the discharge zone. Reactive species had higher average energy and longer residence time in discharge zone at low flux than high flux, however, the number of species were limited, as opposed to high flow rate, which directly affected the performance of plasma on microbial inactivation. For another thing, the E of UV radiation individual changed only within a narrow range, from 0.4 to 1.5, in the whole process no matter how many changes these conditions underwent. These values indicate that during effective bio-decontamination, the contribution of UV radiation to decontaminating features of the oxygen plasma is less. This conclusion is also confirmed by the following.

It is known that UV radiation kills bacilli through destroying the DNA of cells. So, we investigated the state of DNA after oxygen plasma treatment in order to further ascertain the effect of UV radiation on *E. coli*. Figure 3 displays the DNA agarose electrophoresis of *E. coil* after three treatment conditions such as control, UV radiation individual and plasma total treatment at 100 W power, 60 s treatment time and 40 cm^3·min^{-1} oxygen flow rate. As far as we know, if the skeleton of DNA is destroyed to generate small DNA fragments or light molecular DNA when exposes *E. coli* to UV radiation or plasma [22,23], which will produce tail phenomena during the DNA agarose electrophoresis [24]. But the diagrams in Figure 3 (2), which were treated by UV radiation individually, indicate that there are only fewer tail phenomena compared to oxygen plasma total Figure 3 (3). This result makes sure again that UV radiation in oxygen plasma slightly acts on bacilli during bio-decontamination process.

Figure 3. DNA agarose electrophoresis of *E. coil* (1) control; (2) UV radiation individual generated by plasma discharge; (3) plasma total (under 100 W power, 60 s time, 40 cm^3·min^{-1} gas flow rate and 45 Pa pressure).

3.2. Decontamination Effect of Charged Species in Oxygen Plasma

Reactive species strongly bombard the cell surface promoting surface etching. Meanwhile, the living cell cannot quickly repair, resulting in rapid destruction for some microbials in many cases [5]. In addition, charged particles may converge on the surface of cell membranes and cause electrostatic stress when the force is greater than the tensile strength of membrane itself, which will cause damage to cell materials [7,25]. Those action acts on cell materials leading to cell membrane ruptures and leakage of cellular contents, and finally leads the bacterial death [7,11,12]. Therefore, the protein leakage quantity of sample cells was chosen as an index to reflect the inactivation effect of *E. coli* by plasma [24]. Figure 4 shows the tendency of protein leakage quantity in suspensions of bacilli after plasma exposure under different conditions. Compared with the curves of the total decontamination efficiency of oxygen plasma in Figure 2, it can be found that both the changed trends are almost accordant, that is, the action of charged species on the bacilli plays a major role in the inactivation of *E. coli*.

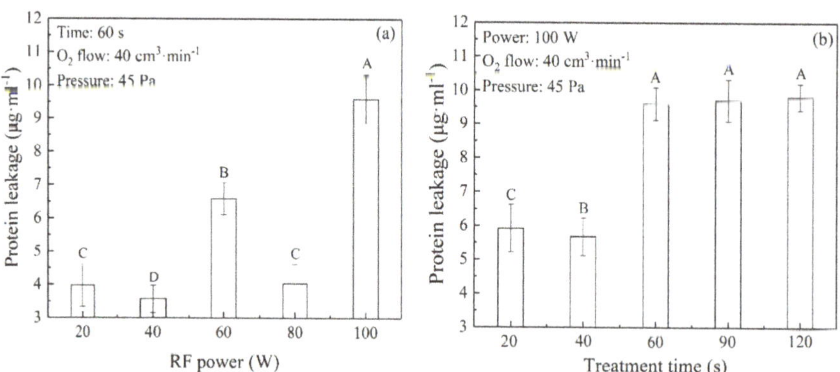

Figure 4. *Cont.*

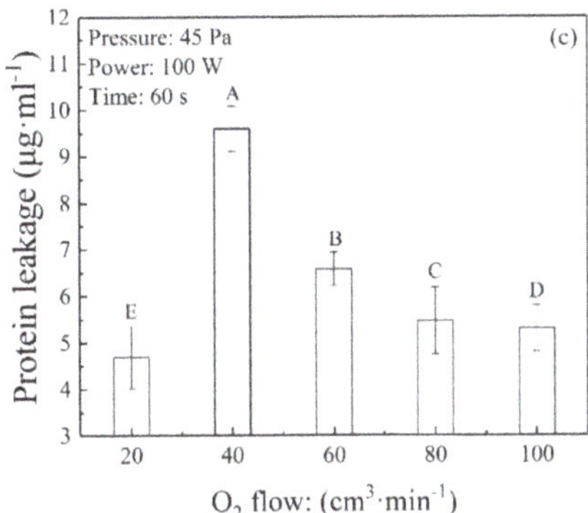

Figure 4. Quantity of protein leakage under different treatment conditions. (**a**) Time 60 s, O$_2$ flow 40 cm^3·min^{-1}, Pressure 45 Pa; (**b**) Power 100 W, O$_2$ flow 40 cm^3·min^{-1}, Pressure 45 Pa; (**c**) Power 100 W, Time 60 s, Pressure 45 Pa. Capital letters A–E represent statistical significance.

This conclusion also can be confirmed further by observing the ultrastructure of bacilli with SEM, as shown in Figure 5. Clearly, the untreated *E. coli* is look like a pole (Figure 5a), whereas after treatment at 100 W power, 60 s treatment time and 40 cm^3·min^{-1} oxygen flow (Figure 5b), bacilli is swollen, incompact and shows obvious 'thawing'. The content of cell leaks completely. This is just due to the intense damage action on cell membrane by the charged particles with high energy.

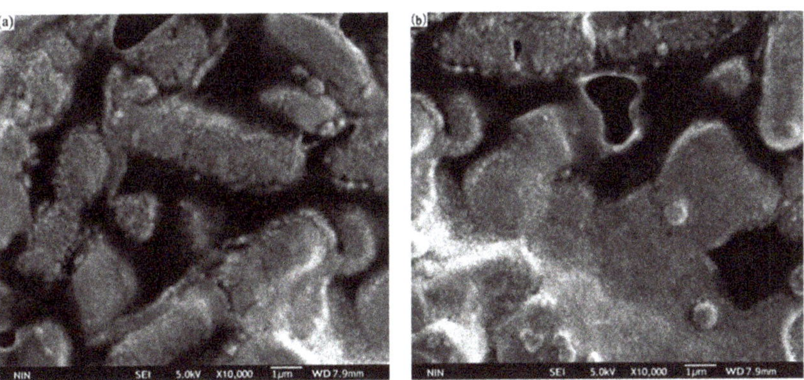

Figure 5. SEM pictures of *E. coil* (**a**) control; (**b**) treated under 100 W power, 60 s time, 40 cm^3·min^{-1} gas flow rate and 45 Pa pressure.

3.3. Decontamination Effect of Free Radicals in Oxygen Plasma

Acting on PUFA in cell membrane by oxygen radicals, specifically highly reactive oxygen atom and hydroxyl radicals, can produce MDA, which is marker as one of the end products of the lipid peroxidation. MDA can destroy the whole structure and functional properties of cell membrane. With the increase of MDA concentration, the content of unsaturated fatty acids in cell membrane reduces, electrical potential of membrane changes and finally, the liquidity of cell membrane is lost [26,27]. So, the changes of MDA concentration in suspensions of bacilli after treatment, as shown in Figure 6, can be used to analyze the interaction between oxygen radicals and cell of bacilli [26,27]. From Figure 6, the MDA concentration has an apparent increase ($p < 0.05$) in a short time (<40 s), and then, little changes occur up to 60 s. Subsequently, it increases gradually again. Also Compared with the curves of the total decontamination efficiency of oxygen plasma in Figure 2, one can find that when treatment time is lower than 40 s, the oxygen radicals in plasma play a major role at a certain extent in the *E. coli* killing process, whereas when the reaction goes on, the contribution of oxygen radicals for inactivation is weakened, whose possible reason is the restraint by intense etching action.

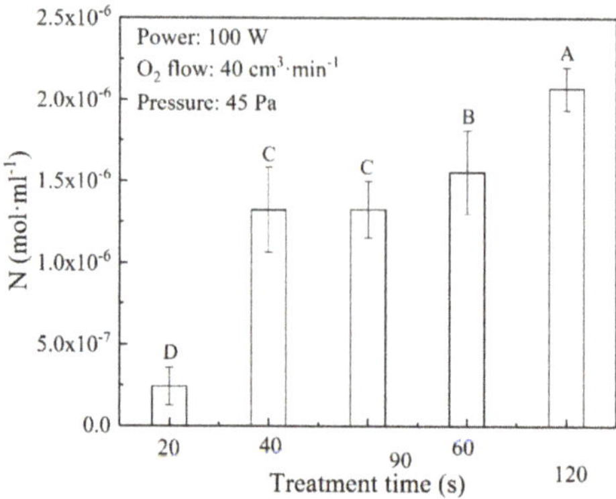

Figure 6. Changes of malondialdehyde (MDA) concentration with increasing the plasma treatment time.

4. Conclusions

In this study, a thorough understanding about the respective contribution of different reactive species for the decontamination of *E. coli* from surfaces in reduced-pressure oxygen plasma was carried out as the following steps: comparing the decontamination efficacy between total plasma and individual UV radiation, measuring the state of DNA and the ultrastructure of bacilli as well as the concentration of protein leakage and MDA production in bacterial suspension after plasma treatment. With the results, the essential effectiveness on *E. coli* of the oxygen plasma was believed to be attributed to the intense etching action of electrons and ions on the bacilli materials. Attacking PUFA in the cell membrane by oxygen radicals plays a major role at a certain extent only during the initial inactivation, and then is restrained by the effect of charged particles with the reaction underway. The function of UV radiation is assistant in the whole process, which results in only the slight damage and rupture of DNA.

Author Contributions: Conceptualization, H.L.; methodology, H.L.; validation, X.F., J.X. and X.M.; formal analysis, X.F.; investigation, X.M.; resources, C.H.; data curation, J.X.; writing—original draft preparation, H.L.; supervision, C.H.; project administration, H.L.

Funding: This work was financially supported by the National Natural Science Foundation of China (No. 31871889), the Natural Science Basic Research Plan in Shaanxi Province of China (No. 2017JM5067), the Fundamental Research Funds for the Central Universities (No. xjj2015132).

Acknowledgments: We also wish to thank Professor Jierong Chen for her support of this work.

Conflicts of Interest: The authors declare no conflict of interest. The funders had no role in the design of the study; in the collection, analyses, or interpretation of data; in the writing of the manuscript, or in the decision to publish the results.

References

1. Moisan, M.; Barbeau, J.; Moreau, S.; Pelletier, J.; Tabrizian, M.; Yahia, L. Low-temperature sterilization using gas plasmas: a review of the experiments and an analysis of the inactivation mechanisms. *Int. J. Pharm.* **2001**, *226*, 1–21. [CrossRef]
2. Lee, T.; Puligundla, P.; Mok, C. Inactivation of foodborne pathogens on the surfaces of different packaging materials using low-pressure air plasma. *Food Control.* **2015**, *51*, 149–155. [CrossRef]
3. Jiang, C.; Schaudinn, C.; Jaramillo, D.E.; Webster, P.; Costerton, J.W. In vitro antimicrobial effect of a cold plasma jet against *Enterococcus faecalis* biofilms. *ISRN Dent.* **2012**, *2012*. [CrossRef] [PubMed]
4. Shi, X.M.; Zhang, G.J.; Yuan, Y.K.; Ma, Y.; Xu, G.M.; Yang, Y. Research on the inactivation effect of low-temperature plasma on Candida albicans. *IEEE Trans. Plasma Sci.* **2008**, *36*, 498–503.

5.   Misra, N.N.; Tiwari, B.K.; Raghavarao, K.S.M.S.; Cullen, P.J. Nonthermal plasma inactivation of food-borne pathogens. *Food Eng. Rev.* **2011**, *3*, 159–170. [CrossRef]
6.   Phan, K.T.K.; Phan, H.T.; Brennan, C.S.; Phimolsiripol, Y. Nonthermal plasma for pesticide and microbial elimination on fruits and vegetables: An overview. *Int. J. Food Sci. Tech.* **2017**, *52*, 2127–2137. [CrossRef]
7.   Bourke, P.; Zuizina, D.; Han, L.; Cullen, P.J.; Gilmore, B.F. Microbiological interactions with cold plasma. *J. Appl. Microbiol.* **2017**, *123*, 308–324. [CrossRef]
8.   Surowsky, B.; Schlüter, O.; Knorr, D. Interactions of non-thermal atmospheric pressure plasma with solid and liquid food systems: A review. *Food Eng. Rev.* **2015**, *7*, 82–108. [CrossRef]
9.   Niemira, B.A. Cold plasma decontamination of foods. *Annu. Rev. Food Sci. Technol.* **2012**, *3*, 125–142. [CrossRef]
10. Lacombe, A.; Niemira, B.A.; Gurtler, J.B.; Fan, X.; Sites, J.; Boyd, G.; Chen, H. Atmospheric cold plasma inactivation of aerobic microorganisms on blueberries and effects on quality attributes. *Food Microbiol.* **2015**, *46*, 479–484. [CrossRef] [PubMed]
11. Dobrynin, D.; Fridman, G.; Friedman, G.; Fridman, A. Physical and biological mechanisms of direct plasma interaction with living tissue. *New J. Phys.* **2009**, *11*, 115020. [CrossRef]
12. Joshi, S.G.; Cooper, M.; Yost, A.; Paff, M.; Ercan, U.K.; Fridman, G.; Friedman, G.; Fridman, A.; Brooks, A.D. Nonthermal dielectric-barrier discharge plasma-induced inactivation involves oxidative DNA damage and membrane lipid peroxidation in Escherichia coli. *Antimicrob. Agents Chemother.* **2011**, *55*, 1053–1062. [CrossRef] [PubMed]
13. Laroussi, M.; Leipold, F. Evaluation of the roles of reactive species, heat, and UV radiation in the inactivation of bacterial cells by air plasmas at atmospheric pressure. *Int. J. Mass Spectrom.* **2004**, *233*, 81–86. [CrossRef]
14. Liu, H.X.; Chen, J.R. Analysis of surface sterilization and properties of medical poly(tetrafluoroethylene) in remote argon plasma. *IEEE Trans. Plasma Sci.* **2008**, *36*, 230–236. [CrossRef]
15. Hertwig, C.; Reineke, K.; Ehlbeck, J.; Knorr, D.; Schlüter, O. Decontamination of whole black pepper using different cold atmospheric pressure plasma applications. *Food Control.* **2015**, *55*, 221–229. [CrossRef]
16. Halfmann, H.; Bibinov, N.; Wunderlich, J.; Awakowicz, P.A. Double inductively coupled plasma for sterilization of medical devices. *J. Phys. D. Appl. Phys.* **2007**, *40*, 41–45. [CrossRef]
17. Hu, M.; Chen, J.R. Inactivation of Escherichia coli and properties of medical poly (vinyl chloride) in remote-oxygen plasma. *Appl. Surf. Sci.* **2009**, *255*, 5690–5697.
18. Yang, L.Q.; Chen, J.R.; Gao, J.L. Low temperature argon plasma sterilization effect on pseudomonas aeruginosa and its mechanisms. *J. Electrostat.* **2009**, *67*, 646–651. [CrossRef]
19. Lerouge, S.; Wertheimer, M.R.; Marchand, R.; Tabrizian, M.; Yahia, L. Effect of gas composition on spore mortality and etching during low-pressure plasma sterilization. *J. Biomed. Mater. Res.* **2000**, *51*, 128–135. [CrossRef]
20. Liu, H.X.; Chen, J.R.; Yang, L.Q.; Zhou, Y. Long-distance oxygen plasma sterilization: Effects and mechanisms. *Appl. Surf. Sci.* **2008**, *254*, 1815–1821.
21. Perez, J.M.; Arenas, F.A.; Pradenas, G.A.; Sandoval, J.M.; Vasquez, C.C. Escherichia coli yqhd exhibits aldehyde reductase activity and protects from the harmful effect of lipid peroxidation-derived aldehydes. *J. Biol. Chem.* **2008**, *283*, 7346–7353. [CrossRef]
22. Sambrook, J.; Russell, D.W. *Condensed Protocols from Molecular Cloning: A Laboratory Manual*; Translated by Huang, P.T.; Science Press: Beijing, China, 2002; pp. 387–396.
23. Mogul, R.; Bol'shakov, A.A.; Chan, S.L.; Stevens, R.M.; Khare, B.N.; Meyyappan, M.; Trent, J.D. Impact of low-temperature plasmas on deinococcus radiodurans and biomolecules. *Biotechnol. Prog.* **2003**, *19*, 776–783. [CrossRef] [PubMed]
24. Gu, C.Y.; Xue, G.B.; Ju, X.C. The study on bactericidal mechanism of plasmas ozone against *E. coli* on the surface. *Mod. Prevent. Med.* **2004**, *31*, 33–35.
25. Arana, I.; Santorum, P.; Muela, A.; Barcina, I. Chlorination and ozonation of wastewater: comparative analysis of efficacy through the effect on Escherichia coli membranes. *J. Appl. Microbial.* **1999**, *86*, 883–888. [CrossRef]

Appl. Sci. **2019**, *9*, 1933

26. Pang, Z.J.; Zhou, M.; Chen, Y. *Study Method of Free Radical Medicine*; Sanitation Press: Beijing, China, 2000; 62p.
27. Köse, K.; Yazici, C.; Cambay, N.; Aşcioğlu, O.; Doğan, P. Lipid peroxidation and erythrocyte antioxidant enzymes in patients with Behcet's disease. *Tohoku J. Exp. Med.* **2002**, *197*, 9–16. [CrossRef] [PubMed]

Article

Modulation of Metamorphic and Regenerative Events by Cold Atmospheric Pressure Plasma Exposure in Tadpoles, *Xenopus laevis*

Ma Veronica Holganza [1,†], Adonis Rivie [2], Kevin Martus [3,†] and and Jaishri Menon [1,*,†]

1   Department of Biology, William Paterson University, Wayne, NJ 07470, USA
2   Department of Integrative Biology, University of California Berkeley, Berkeley, CA 94720, USA
3   Department of Physics, William Paterson University, Wayne, NJ 07470, USA
*   Correspondence: menonj@wpunj.edu
†   These authors contributed equally to this work.

Received: 3 June 2019; Accepted: 14 July 2019; Published: 18 July 2019

Abstract: Atmospheric pressure plasma has found wide clinical applications including wound healing, tissue regeneration, sterilization, and cancer treatment. Here, we have investigated its effect on developmental processes like metamorphosis and tail regeneration in tadpoles. Plasma exposure hastens the process of tail regeneration but delays metamorphic development. The observed differences in these two developmental processes following plasma exposure are indicative of physiological costs associated with developmental plasticity for their survival. Ultrastructural changes in epidermis and mitochondria in response to the stress of tail amputation and plasma exposure show characteristics of cellular hypoxia and oxidative stress. Mitochondria show morphological changes such as swelling with wide and fewer cristae and seem to undergo processes such as fission and fusion. Complex interactions between calcium, peroxisomes, mitochondria and their pore transition pathways are responsible for changes in mitochondrial structure and function, suggesting the subcellular site of action of plasma in this system.

Keywords: atmospheric pressure plasma; developmental plasticity; metamorphosis; mitochondria; regeneration; tadpoles; ultrastructure

---

1. Introduction

Advances and interactions between scientific fields such as biology, physics, and medicine have brought many benefits in tissue engineering and regenerative medicine. Plasma medicine, combining physics of plasma with medicine, has developed rapidly in the last few years [1] and it is the subject of this broad interdisciplinary research.

1.1. Plasma Properties

An ionized gas is an ensemble of positively and negatively charged particles. The term plasma was applied to these systems by Langmuir [2] and should not be confused with blood plasma. Over the past twenty years, atmospheric pressure plasmas have been the subject of numerous reviews regarding the state of the art of the devices used to generate these plasmas and characteristics of the plasmas [3–6]. In the present study, a plasma jet was used to generate a cold (low-temperature) non-equilibrium (non-thermal) atmospheric pressure plasma and will be designated as APP.

APP is a relatively new approach that is extensively studied for its biomedical applications [7]. These include sterilization, blood coagulation, cancer treatment, cell metabolism modification, hospital hygiene, antifungal treatment, dental care, skin disease and wound healing [8–13]. Tissue engineering and regenerative process involves a variety of cellular activities such as cell proliferation, differentiation and apoptosis. Exposure to APP promotes repair of muscle defects through control of cell proliferation and differentiation, suggesting that APP exposure has the potential for use in muscle tissue engineering and regenerative therapies [14]. APP exposure has been shown to induce differentiation and growth of developing mouse limb buds [15], as well as differentiation of bone marrow stem cells [16].

1.2. Plasma and Reactive Oxygen Species and Reactive Nitrogen Species

Plasma generated from atmospheric air includes reactive species mainly from N_2 and O_2, and may contain N, O, O_3, NO, NO_2, and OH. The final products of the species present in the plasma are several Reactive Oxygen Species (ROS) and Reactive Nitrogen Species (RNS). These reactive molecules expose biological material to oxidation processes [17–20]. The reactive species are generated either within the plasma itself or in the tissue when brought in contact with plasma [21]. Both ROS and RNS are also derived within the cell from organelles such as mitochondria, peroxisomes, and the endoplasmic reticulum.

ROS and RNS have pleiotropic effects on cellular physiology, which include metabolism, cellular signaling and morphogenetic processes [22,23]. At moderate concentrations, beneficial effects of ROS/RNS include defense against infectious agents, activating cellular signaling pathways and mitogenic response [24]. However, overproduction of ROS/RNS in cells or tissue can result in oxidative stress in the microenvironment of the cells [20], which can damage cell structures including lipids and membranes, proteins, and DNA [25] triggering cell death by apoptosis or necrosis [26,27]. It was also reported that exposure to plasma leads to different behaviors in cells in vitro, such as decrease in cell migration, cell detachment, apoptosis or necrosis depending on power and the exposure time [8]. APP exposure produces dose-dependent effects that range from increased cell proliferation to apoptosis including DNA damage, oxidative stress and cell cycle modification [26,28–32]. Previous studies on tadpole tail regeneration from our lab have shown that the longer exposure of the amputated tail of tadpoles *Xenopus laevis* to APP resulted in enhanced mortality, ascribed to increased production of ROS [33].

1.3. Plasma and Wound Healing in Mammals and Amphibians

APP exposure is also a tool that can promote wound healing and epithelial regeneration which are dynamic and complex processes involving many different types of cells. APP exposure is known to improve wound healing by its antiseptic effect, stimulation of proliferation and migration of wound related skin cells, activation or inhibition of integrin receptors on the cell surface or its angiogenic effect [34]. Appropriate doses of cold plasma have antibacterial effects, activate fibroblast proliferation in wound tissue and thus promote wound healing in mice [35]. Interaction of ROS and RNS with wounded tissue accelerates repair processes without any adverse effects on normal tissue [36].

However, mechanisms that cause adult/postnatal mammalian skin healing differ from embryonic (fetal) skin development and repair as well as from organisms such as amphibians, which regenerate their injured tissue in a process analogous to development [37,38]. There is no scar formation in embryonic mammalian skin [39], as well as following tail amputation in amphibians [40] in contrast to adult mammalian skin. Anuran amphibians, such as *Xenopus laevis* in its larval stage, can regenerate their limb and tail by means of the formation of blastema which is a mass of dedifferentiated, proliferating cells [41]. This remarkable ability of amphibians makes them a very attractive model for the study of regeneration.

1.4. Cell Organelles, Oxidative Stress, and Apoptosis

Mitochondria and peroxisomes are dynamic organelles that have intricate structural and functional relationships [42,43]. Both of these organelles play a key role in ROS production, which is important for cell signaling and ROS scavenging [44]. They also show high plasticity and easily adapt in response to developmental, metabolic and environmental alterations [42]. Peroxisomal function has long been linked to oxygen metabolism due to their high concentration of H_2O_2 generating oxidases and have several antioxidant systems, such as catalase, superoxide dismutase, peroxisomal membrane protein 20 and glutathione peroxidase [45]. Mitochondria also possess enzymes and non-enzymatic antioxidant systems [46]. Mitochondria are quantitatively the most important source of ROS, which maintain the steady-state concentration at non-toxic levels by a variety of antioxidant defenses and repair enzymes [47]. Oxidative stress is an expression used to describe various deleterious processes resulting from an imbalance between the excessive formation of ROS and limited antioxidant defenses.

Oxidative stress associated with cell death in the developing limb is due to mitochondrial production of ROS [48]. Mitochondria play a key role in apoptotic processes. The mitochondrial permeability transition pore (mPTP) is a large conductance channel that opens in the mitochondrial membrane in response to high calcium (Ca^{2+}), low ATP, and oxidative stress [49–52], and opening of the mPTP causes abrupt mitochondrial depolarization. APP treatment is associated with an alteration in morphology of mitochondria, a decrease in mitochondrial membrane potential, mitochondrial enzyme activity and respiration rate in cancer cells [18,53].

Ca^{2+}, one of the most ubiquitous cellular second messengers, is also known to be involved in cell death by apoptosis, necrosis and autophagy [54]. Mitochondria are the paradigm of the double-edged sword effect of Ca^{2+} on cell survival and death [55]. Ca^{2+} overload could be detrimental to the cells as it can lead to mitochondrial depolarization, cytochrome c release, lipid peroxidation, transcription factor activation and DNA damage, resulting in apoptotic and non-apoptotic cell death [56]. Peroxisomes also play an important role in Ca^{2+} homeostasis.

1.5. Developmental Plasticity in Tadpoles

Amphibians are a group of vertebrates with a great range of developmental plasticity in their life history. Many of them maintain an aquatic larval stage of life as tadpoles that metamorphose into a terrestrial one. These organisms are also known to show adaptive developmental plasticity in response to environmental changes [57,58], such as ability for accelerated metamorphosis in drying ponds. This rapid growth comes at a cost, and one such suggested cost of compensatory growth is oxidative stress due to increased ROS [59]. Environmental assessment by amphibian larvae is so well regulated that developmental acceleration can even decelerate if the conditions in the aquatic system ameliorate [58].

1.6. Organ Regeneration in Amphibians

The restoration of lost or damaged body parts continues to be a topic of great interest for biologists. The ability to regenerate body parts varies greatly from one species to another. This feature is extensively studied in amphibian tadpoles of the species *Xenopus laevis* and *Xenopus tropicalis*, which have remarkable abilities to regenerate their tails and limbs following amputation [60–62]. Unlike most amphibians which undergo metamorphosis, there are some species like the axolotl, which retains larval features such as external gills and undeveloped limbs throughout their life—a condition referred to as neoteny. These organisms are aquatic and fully mature but fail to metamorphose and are capable of regenerating wounds on their skin with scarless healing [40]. Increased regenerative ability of amphibians makes them excellent lab models for biomedical research, such as traumatic injury.

Most of the studies on effects of plasma are focused on tissue regeneration. The main purpose of our study is to explore the effect of plasma on organ regeneration, such as tail replacement, following amputation. Additionally, we also studied the effect of plasma on post-embryonic development, such as metamorphosis, to get insights into adaptive plasticity in these organisms to cope with oxidative stress following plasma exposure.

2. Materials and Methods

2.1. Plasma Discharge Source

Figure 1 illustrates the geometry of the APP discharge source. The flow of high purity Helium gas (Airgas Zero Grade) was controlled over a range of 10 to 150 standard cubic centimeters per minute (sccm) by a (Cole Parmer PMR1-010537, Chicago, IL, USA) gas flowmeter and a standardized rate of 50 sccm was used for all trials. The gas flowed through a 15 cm long quartz tube with an outer diameter of 6.35 mm. A copper electrode was attached to the outer surface of the tube at a point that was 5 cm from the end of the tube. An alternating current power supply operating at 15 kHz and 17 kV was attached to the electrode. The power supply's operating parameters were controlled such that the discharge was restricted to the flow region between the electrode and the exit aperture of the tube. The resulting non-equilibrium plasma had an emission profile that indicated that OH radicals and molecular nitrogen was present within the tube.

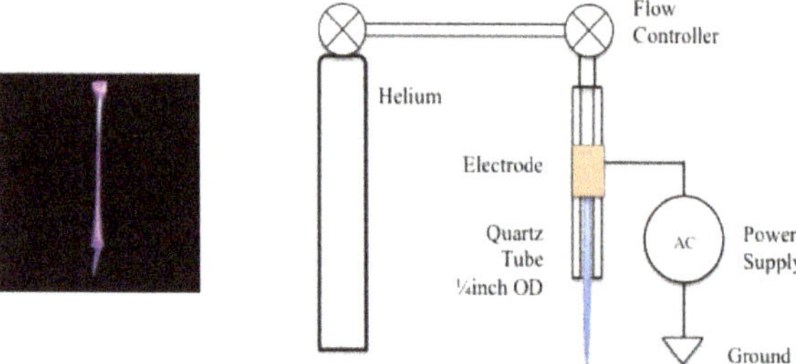

Figure 1. The plasma discharge was generated using a gas flow through a 6.35 mm outer diameter quartz tube with a single electrode powered by an alternating current power supply. The power supply used had a frequency of 18 kHz and a 17 kV peak to peak voltage. Helium was flowed through the tube at a rate of 50 sccm. The tadpoles were exposed indirectly to the glow region of the discharge, but approximately 3 cm beneath the tube. Inside the tube, optical emission spectroscopy confirmed that the emissions were produced by OH radicals and molecular nitrogen.

2.2. Maintenance of Tadpoles

Xenopus laevis tadpoles, purchased from Xenopus I (Ann Arbor, MI, USA) were maintained in the laboratory in aquaria and fed on a standard diet obtained from the same company. Water in the aquaria was changed on every alternate day. For the present study, the tadpoles at stage 57 were used [63] (see Appendix A for a description of stages).

2.3. Study of Metamorphosis

Tadpoles at stage 57 were divided in three groups of 20 each. Amputation of tail (40% removal) and plasma exposure for 40 s was carried out by anesthetizing tadpoles on ice in the groups assigned below:

- Group 1: Amputated and treated with plasma and this group will be referred as experimental in the text.
- Group 2: Amputated and not treated which will be referred as control in the text.
- Group 3: This group of tadpoles were not amputated or treated with plasma to see the progress of metamorphosis which served as the control for group 2.

All the three groups were monitored for metamorphic development according to stages described by Nieukoop and Faber [63].

2.4. Tail Regeneration Study

Tadpoles at stage 57 were divided into the following two groups:

- Group 1: Tadpoles amputated and exposed to plasma as mentioned above (experimental).
- Group 2: Tadpoles amputated but not exposed to plasma (control).

For the following studies, tadpoles were anesthetized in MS222 and wound epithelium at 24 h and blastema at 5 days were removed from both of the groups.

2.5. Ca^{2+} Quantification Assay

Ca^{2+} assay was carried out using an Abcam calcium assay kit (ac102505) and quantified through microplate photometry. Wound epithelium and blastema for both of the groups were pooled from five tadpoles, as tissue from an individual tadpole was not enough for this assay (due to small size of the organism). The experiment was repeated four times ($n = 4$).

2.6. Confocal Microscopy

All the kits for in situ staining Calcium (F10489), peroxisomes (S34203) and mPTP (I35103) were purchased from Molecular Probes (Thermoscientific, Chicago, IL, USA). For Ca^{2+} staining, Fluo-4 dye is used which binds to Ca^{2+} exhibiting green fluorescence. mPTP staining was carried out using a calcium/cobalt quenching technique which uses calcein acetoxymethyl that enters the cell followed by treatment with cobalt chloride to quench cytosolic calcein fluorescence (green). Mitochondria are stained with Mito Tracker dye that appears red. If mPTP is compromised, calcein enters mitochondria but not the quencher. Thus, green fluorescence of the dye overlays red mitochondria producing yellow color. Peroxisome staining was carried out using antibody directed against peroxisomal membrane protein 70 (PMP70).

Tissues were harvested at 24 h and 5 days. For confocal imaging studies of Ca^{2+}, mPTP and peroxisomes, three tadpoles were used for each staining procedure ($n = 3$) and carried out according to the method provided in the staining kit. Tissue samples were visualized under a Zeiss LSM 510 META Confocal microscope (Peabody, MA, USA).

2.7. Transmission Electron Microscopy

Tissue samples at 24 h and 5 days from control and experimental tadpoles ($n = 3$ per time point) were fixed in 2.5% glutaraldehyde in 0.1 M cacodyalte buffer for 24 h, washed in the same buffer, osmicated in 1% osmium tetroxide (OsO_4), dehydrated, through a graded series of alcohol, and routinely embedded in Epon 812. Ultrathin sections in the silver gray range were cut and stained with uranyl acetate and lead citrate and observed under Hitachi 7700 (Hitachi High Technologies America, Pleasanton, CA, USA).

3. Results

3.1. Metamorphic Studies

Following tail amputation, APP exposure delayed metamorphosis compared to non-treated, see Figure 2. In group 1 (amputated and treated), all tadpoles metamorphosed by day 36 compared to group 2 (amputated and not treated), where by day 17 all of them turned into froglets. At day 10, there was a significant change between experimental and control tadpoles. There was no significant difference in the progress of metamorphosis between group 2 (amputated and not treated) and group 3 (not amputated or treated).

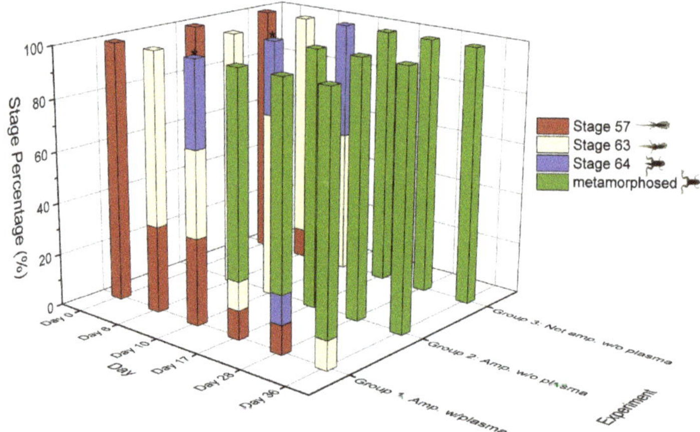

Figure 2. Graphic representation of developmental stages in metamorphosis impacted by tail amputation and plasma exposure. See Appendix A for details of developmental stages (57, 63, 64). At stage 57, three groups of tadpoles were divided as follows: Group 1: amputated and treated with plasma (experimental tadpoles). Group 2: amputated and not treated to plasma (control tadpoles). Group 3: not amputated and not treated to plasma. At day 6, in group 1, 40% of tadpoles were at initial stage 57 and the rest had progressed to stage 63, whereas, in group 2, all the tadpoles were at stage 63 and in group 3, 90% of tadpoles were at stage 63. At day 10 in group 1, approximately equal number of tadpoles (30 to 33%) were at stage 57, 63 and 64 compared to group 2 and 3, where approximately 80% of tadpoles were at stage 63 and the rest at stage 64. By day 17, in group 1, an equal percentage (15%) were at stage 57 and 63 and the rest had developed into froglets compared to groups 2 and 3 where all the tadpoles had metamorphosed into froglet. By day 36, in group 1, 90% of the tadpoles had metamorphosed into froglets.

Once the tail was amputated at stage 57 (Figure 3a), wound healing was completed by 24 h and a wound epithelium was formed (Figure 3c) in both the experimental and control group, subsequent to which a mass of dedifferentiated cells accumulated under the wound epithelium and by the 5th day a structure called a blastema was formed (Figure 3d,e).

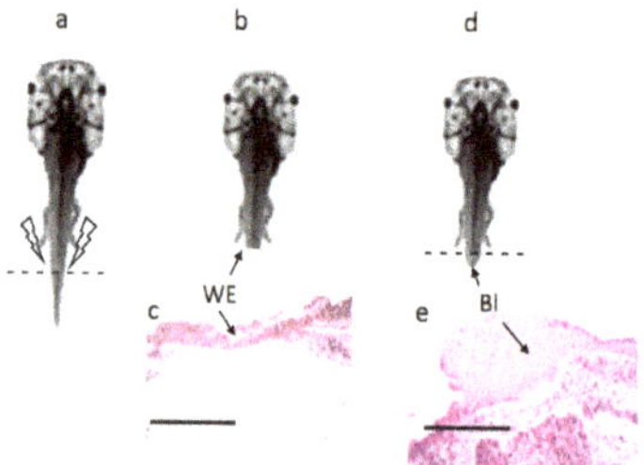

Figure 3. Wound healing and blastema formation: Following tail amputation at stage 57 (Figure 3a), wound was healed and two to three layers wound epithelium (WE) was formed in both the groups by 24 h (Figure 3b,c). By the 5th day, a mass of dedifferentiated cells was formed called blastema (BL) in both the groups (Figure 3d,e); scale bar ——— 100 µm. However, the size of the blastema was slightly larger in the experimental group compared to the control [33].

3.2. Ca^{2+}

In situ staining showed that, in wound epithelium at 24 h, fluorescence intensity for Ca^{2+} was higher (Figure 4a) compared to control (Figure 4b). In blastema, the fluorescence intensity for Ca^{2+} had reduced in experimental tadpoles (Figure 4c) in comparison with control (Figure 4d).

Quantitative analysis of 24 h and day 5 blastema from experimental and control tadpoles showed similar results (Figure 5). However, statistically, the differences were not significant. This could be explained by the fact that, under microscope, wound epithelium and blastemal are clearly discernible from the underlying tissues. When samples were pooled from five tadpoles for quantitative analysis, it is inevitable that some of the underlying tissue like muscles would contribute to total quantity of Ca^{2+}.

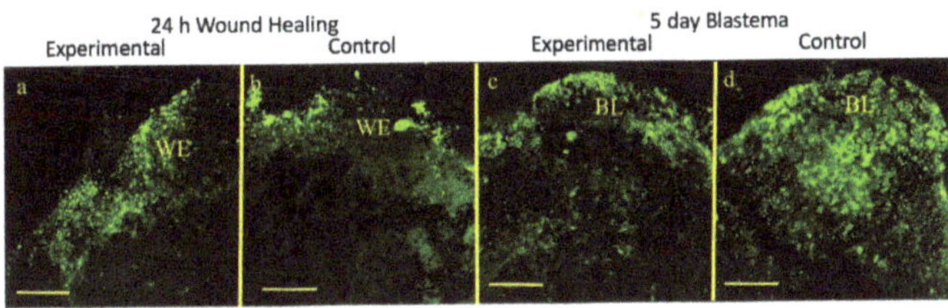

Figure 4. Ca^{2+} fluorescence staining: At 24 h post amputation, wound epithelium (WE) of experimental tadpoles show higher intensity of Ca^{2+} fluorescence (green) (Figure 4a) compared to control (Figure 4b). However, in day 5 blastema (BL) of experimental tadpoles, fluorescent staining was lower in Figure 4c compared to control Figure 4d; scale bar ——— 100 μm.

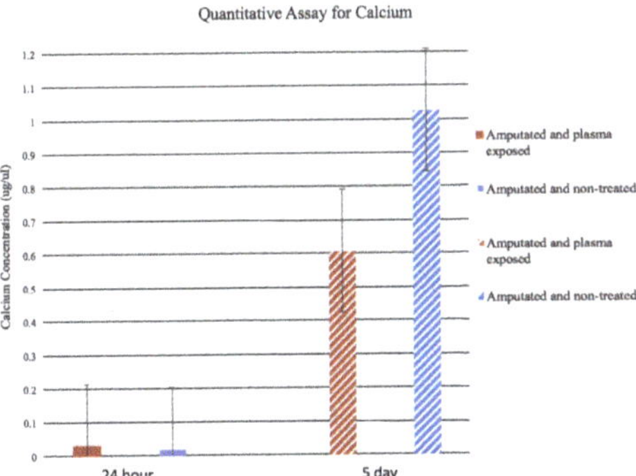

Figure 5. Quantitative Ca^{2+} estimation: At 24 h post amputation, wound epithelium (WE) of experimental and control tadpoles show lower Ca^{2+} content, whereas, in day 5, blastema Ca^{2+} content in both groups was much higher. The changes were insignificant between the two groups.

3.3. mPTP

In experimental tadpoles, at 24 h, mitochondrial staining was prominent (Figure 6a) compared to control (Figure 6b). Similarly, the number of mitochondria and opened pores in the mitochondrial membrane of the blastema in experimental tadpoles were very distinct (Figure 6c) compared to control where there is no staining of mPTP observed (Figure 6d).

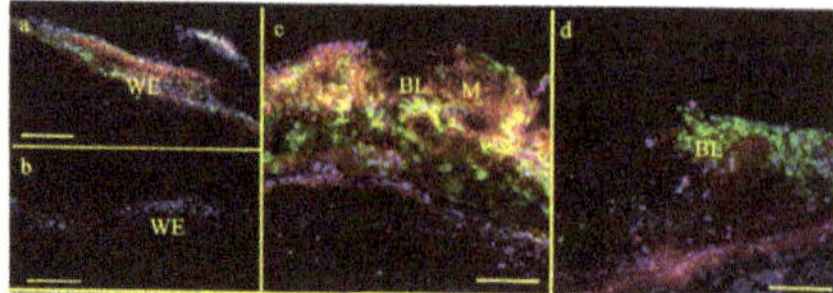

Figure 6. mPTP staining: Wound epithelium (WE) at 24 h as well as 5 day blastema (BL) of experimental tadpoles show increased staining for mitochondria as well as open mitochondrial pores (Figure 6a,c) compared to control (Figure 6b,d). Mitochondria (M)—red; mPTP open—yellow (calcein has entered mitochondria); calcein—green (in cytoplasm); nuclei—blue; scale bar ——— 100 µm.

3.4. Peroxisomes

Staining of peroxisomes in blastema of experimental tadpoles is significant (Figure 7a) compared to control, where there was negligible staining for these organelles (Figure 7b).

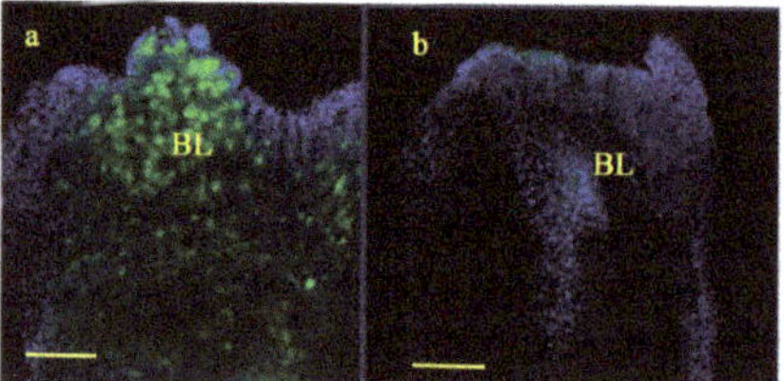

Figure 7. Increased peroxisomal staining (green) is seen in experimental tadpoles (Figure 7a) compared to control (Figure 7b). blue—nuclei. Scale bar ——— 100 µm.

3.5. Light and Electron Microscopy

At 24 h, wound epithelium is formed in both the experimental and the control tadpoles (Figure 8a,b). However, there is an increased number of lipid inclusions (Figures 8a and 9b) and decreased visible intercellular spaces in experimental (Figures 8a and 9a) compared to control tadpoles (Figures 8b and 9c). At 5 days post amputation, plasma treated tadpole tail also displayed swelling of cells and blebbing of the plasma membrane (Figure 8c,e) compared to control (Figure 8d,f).

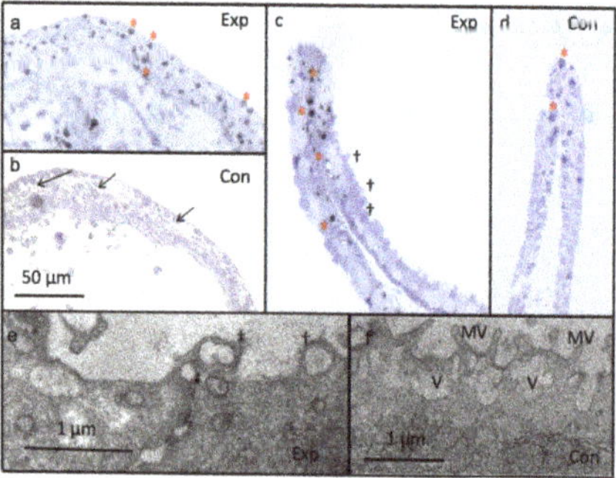

Figure 8. 1 to 1.5 µm thick plastic sections showing fully formed wound epithelium after 24 h post amputation in experimental (Figure 8a) as well as control groups (Figure 8b). Note osmiophilic lipid bodies (red asterisk *), and decreased intercellular spaces in experimental tadpoles (Figure 8a) compared to control where intercellular spaces are visible (←—) (Figure 8b). Note blebbing (†) of the cells in epithelium of blastema (5 days) of experimental tadpoles (Figure 8c) compared to control (Figure 8d). Ultrastructural studies; Figure 8e,f: Transmission electron micrograph of wound epithelium of experimental tadpoles shows loss of microvilli and blebbing (†) (Figure 8e) in contrast to control (Figure 8f). (Figure 8a–d from Rivie et al. [33]). Exp-experimental; Con-control.

At 24 h post amputation, tail epidermis of experimental tadpoles revealed loss of microvilli (Figure 9a,b) compared to control (Figure 9c,d). Mitochondria of wound epithelium at 24 h of the experimental tadpoles shows loss of cristae (Figure 9e) and fusion of mitochondria with wide cristae (Figure 10a,b) compared to control where mitochondria are small and round with thin cristae (Figure 9d,f). Additionally, in experimental tadpoles, the mitochondrial matrix is electro-lucent (Figure 10a,c) compared to control (Figure 10d). Close proximity of mitochondria with endoplasmic reticulum is seen in experimental (Figure 10c) and control (Figure 10d) tadpoles.

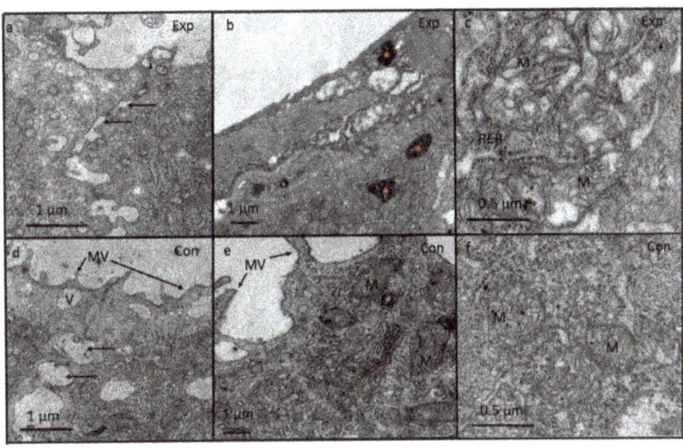

Figure 9. Ultrastructure of wound epithelium at 24 h post amputation showing loss of microvilli (MV), decreased intercellular spaces (←—) (Figure 9a) and increased number of lipid inclusions (*) (Figure 9b) compared to control where microvilli (MV) and intercellular spaces (←—) are prominent (Figure 9c,d). In experimental tadpoles, mitochondria are swollen and fused with wide cristae (Figure 9e) compared to control (Figure 9f). Exp-experimental; Con-control.

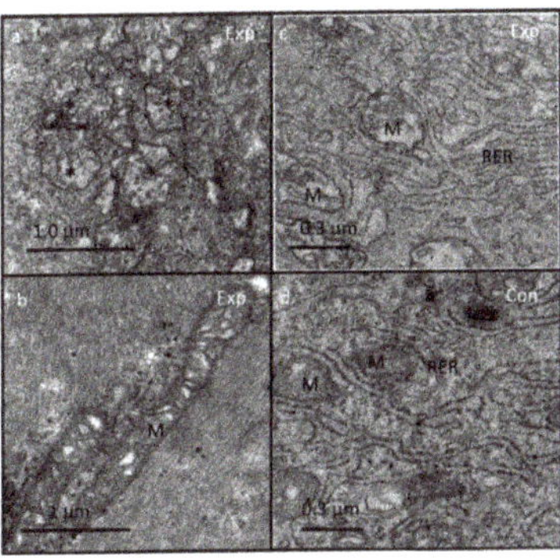

Figure 10. Ultrastructure of blastema from experimental tadpoles after 5 days showing fusion of damaged mitochondria (M) with near loss of cristae (Figure 10a) and often very long mitochondria due to fusion (Figure 10b). Mitochondria (M) from plasma-exposed tadpoles have an electro-lucent matrix (Figure 10c) compared to control (Figure 10d). Note close proximity of endoplasmic reticulum (RER) and mitochondria in both of the groups (Figure 10c,d). Exp-experimental; Con-control.

4. Discussion

4.1. Adaptive Plasticity

Earlier studies [33] and current results indicate that the plasma treatment accelerates tail regeneration of tadpoles *Xenopus laevis* while slowing down the metamorphic progress. This is an example of adaptive plasticity, essential for many species to cope with changes in the environment.

Many amphibian species have exploited this trait by accelerating metamorphic events when threatened by desiccation in their aquatic habitat [58,64]. Increased oxidative stress is considered a key determinant factor for accelerated growth and metamorphosis [65]. The most important environmental adaptation for tadpoles is swimming in water for their survival [66]. In the current study, the tadpole tail was amputated and exposed to plasma at a stage when hind limbs had not yet fully developed and forelimbs had not erupted (stage 57). Hence, these tadpoles are subjected to two kinds of stress: (a) removal of the tail hampers their locomotor ability (movement in water) and (b) the stress of plasma exposure. Restriction of swimming activity in the absence of fully developed limbs contributes to a faster rate of tail regeneration for their survival but is achieved at the expense of metamorphosis. One can argue that these tadpoles can hasten the process of metamorphosis instead of regenerating their tail. However, accelerated metamorphosis cannot take place until the animal has reached a threshold developmental stage [67,68]. Cost of faster growth of the regenerating tail in these tadpoles following plasma exposure is associated with increased oxidative stress trade-offs. Differences observed in tail regeneration and metamorphosis between experimental and control tadpoles indicate the physiological cost involving metabolic machinery at the cellular and organelle level. However, in the absence of plasma treatment, there was no significant difference in metamorphic growth between amputated and non-amputated group of tadpoles. This means that plasma exposure might be increasing the signaling for tail regeneration via bone morphogenetic protein and Notch, which act as mitogenic factors in *Xenopus* organ regeneration as reported by Beck et al. [62]. This topic remains to be explored following plasma treatment.

4.2. Tail Regeneration

The tail of *Xenopus* tadpole is efficient in regenerating all the competent tissues after amputation, including spinal cord, notochord, muscles and epidermis. Upon amputation of the tail, the wound is healed within 24 h and a population of progenitor cells are produced known as the blastema. Blastema is a mass of dedifferentiated pluripotent cells that is able to form multiple tissue types following the amputation and wound healing occurring [69–71]. The process of tadpole tail regeneration is considered analogous to tissue renewal in mammals [72]. Analogies have also been drawn between cells undergoing regeneration in salamanders and stem cells [73].

Our previous and current observations have shown that plasma exposure accelerates the process of regeneration in tadpoles, *Xenopus laevis*, and it was interesting that, under APP exposure, local oxidative stress was higher compared to control [33]. Tadpoles' capacity to adaptively tune to tail regeneration is controlled by several factors—one of them being oxidative stress. Moderate levels of ROS act as signaling molecules during tail regeneration in *Xenopus* [61,74]. Wounding tissue generates ROS especially H_2O_2 [75], which freely diffuses between cells and acts as a signaling molecule during early phases of regeneration and blastema formation in adult fin regeneration of zebrafish [76].

4.3. Ca^{2+}, Endoplasmic Reticulum (ER), Mitochondria and Peroxisomes

Ca^{2+} is an important second messenger and plays a critical role in the fate of the cell surviving or dying. Currently, we have an observed increase in Ca^{2+} with confocal microscopy during wound healing (24 h post amputation) but a decrease in day 5 blastema. It is possible that increased Ca^{2+} content derived from intracellular Ca^{2+} stores, such as ER, following plasma exposure might be responsible for apoptosis of damaged cells during the wound healing period. Low Ca^{2+} in the blastema of experimental tadpoles may support early differentiation of these cells as reported in mammalian epidermal cells [77], resulting in faster tail regeneration in experimental tadpoles.

Additionally, the relationship between Ca^{2+} and ROS is complicated and close, either as a crucial partner in regulating the redox status of cells, determining cell fate, or in signaling in response to a number of physiological and stress conditions [78]. Our studies find support from the work of Ma et al. [29] that an increase in intracellular ROS [33] and Ca^{2+} following APP exposure (current study) could serve as valuable biomarkers for the oxidative stress.

Mitochondria also undergo remodeling of their shape and morphology during programmed cell death—a biological process that usually confers an advantage during an organism's life-cycle. Mitochondria are dynamic organelles that continually undergo fusion and fission. These are two opposing processes that act in concert to maintain the shape, size, and number of mitochondria as well as their physiological function [79]. Our electron microscopy study shows that some mitochondria from experimental tadpoles had fused acquiring elongated or giant shapes with wide cristae compared to controls where mitochondria were round, smaller and with thin cristae. A high level of stress leads to fission of mitochondria, which would eliminate damaged mitochondria through mitophagy [80]. Mitochondrial fission also inhibits propagation of the Ca^{2+} signal and protects against Ca^{2+} mediated cell death [81], whereas mitochondrial fusion allows them to compensate for one another's defects by sharing components, in order to maintain energy output in stressful situations [82,83]. Fission and fusion interact to create a diversity of mitochondrial structures and ER may play an active role in determining the sites of mitochondrial fission [84]. Additionally, in experimental tadpoles, some mitochondria were swollen with an electro-lucent matrix compared to control, which could be due to ischemia [85], increased Ca^{2+} overload [18,55], and induction of mPTP [86]. Increase in Ca^{2+} (as observed using the confocal microscopy), as well as oxidative stress [33] in the regenerated tissue following plasma exposure, might be responsible for the opening of mPTP, which changes mitochondrial morphology as reported by Fridman and Lowe [87]. Currently observed changes in mitochondrial dynamics may allow cells to adapt to oxidative stress which warrants further studies.

ER is also known to generate peroxisomes [88]. Currently, an observed increase in peroxisomes in blastema of experimental tadpoles might be generating and releasing important signal molecules; such as, O_2^-, H_2O_2 and NO in cytosol, which can contribute to a more integrated communication system among cell compartments as reported by Corpas et al. [89]. Additionally, peroxisomes must be providing protective mechanisms to counteract oxidative stress.

4.4. Ultrastructure of Tadpole Epidermis (Epidermal Morphology and Junctional Complex)

The epidermis of vertebrates is the principal barrier against environmental stress. Higher vertebrates have specific detoxifying molecules and enzymes involved in providing protection to the organisms. The importance of tight junctions in barrier formation in mammals is very well documented [90]. However, amphibian epidermis does not have as good a barrier mechanism as in higher vertebrates. However, they do possess complex junctional systems that help to decrease trans-epidermal water loss [91] and maintain skin integrity by preventing toxic, infectious or traumatic stress [92]. In the current study, we have observed adherens junctions (possibly including tight junctions) between epidermal cells of the tail and reduction of intercellular spaces at 24 h post amputation and following plasma exposure. Cellular junctions do not function independently and it is reported that junctional crosstalk, interdependencies, and compensation are necessary for tissue strength [93]. Since tadpoles do not have a good barrier on skin surface, this means that the tension is likely to be in the underlying epidermis. Additionally, the outer surface of the epidermis is exposed to the external environment, and the animal has to prevent the loss of molecules from the cells, which probably resulted in the loss of microvilli and formation of tight junctions between the epidermal cells following plasma exposure. These morphological changes in the epidermis following plasma exposure indicate adaptive changes in order to maintain skin integrity.

5. Conclusions

APP exposure affects wound healing and tail regeneration in larval tadpoles. The APP exposure stimulates multiple signaling cellular pathways coordinating the morphogenetic processes that underlie regeneration and metamorphosis, which in turn influences developmental plasticity of these organisms. Currently, we are extending this work, evaluating gene expression for antioxidant enzymes superoxide dismutase and catalase, as well as bone morphogenetic protein, a factor required for regeneration and metamorphosis.

Author Contributions: K.M. and J.M. conceptualized the idea. K.M. designed and tested the plasma source. J.M. planned the experiments. M.V.H. and A.R. carried out the experiments on the confocal microscope. M.V.H. analyzed and processed the experimental results and contributed to the interpretation of results. J.M. carried out the ultrastructural studies. J.M. and K.M. took the lead in writing the manuscript in consultation with M.V.H.

Funding: This research received no external funding.

Acknowledgments: K.M. and J.M. support through the William Paterson University Assigned Released Program and Student Worker Fund. M.V.H. and A.R. supported through the Garden State Louis Stokes Alliance for Minority Participation (GS-LSAMP). Figure 8a–d are reprinted by permission from Springer Nature, The European Physical Journal—Special Topics, Atmospheric pressure plasma accelerates tail regeneration in tadpoles *Xenopus laevis*, Rivie, Martus, Menon (2017) [33].

Conflicts of Interest: The authors declare no conflict of interest.

Abbreviations

The following abbreviations are used in this manuscript:

APP	Atmospheric Pressure Plasma
BL	Blastema
ER	Endoplasmic Reticulum
M	Mitochondria
mPTP	mitochondrial Permeability Transition Pore
MV	Microvilli
ROS	Reactive Oxygen Species
RNS	Reactive Nitrogen Species
RER	Rough Endoplasmic Reticulum
sccm	standard cubic centimeters per minute
WE	Wound Epithelium

Appendix A

Tadpole development by stages from Nieukoop and Faber [63]. Stage 57: Lateral view of the tadpoles showing hind limbs (←—) in early developmental stage and forelimbs have not erupted. Stage 63: Dorsal view of the tadpole showing well developed hind limbs and erupted forelimbs. Stage 64: Dorsal view of tadpole tail regression.

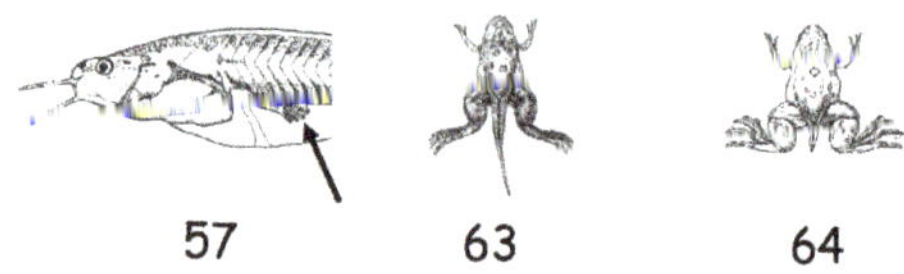

Figure A1. Tadpole development by stages. Hind limbs stage 57 denoted by ←—.

References

1. von Woedtke, T.; Metelmann, H.R.; Weltmann, K.D. Clinical plasma medicine: State and perspectives of in vivo application of cold atmospheric plasma. *Contrib. Plasma Phys.* **2014**, *54*, 104–117. [CrossRef]
2. Langmuir, I. Oscillations in ionized gases. *Proc. Natl. Acad. Sci. USA* **1928**, *14*, 627–637. [CrossRef] [PubMed]
3. Schütze, A.; Jeong, J.Y.; Babayan, S.E.; Park, J.; Selwyn, G.S.; Hicks, R.F. The atmospheric-pressure plasma jet: A review and comparison to other plasma sources. *IEEE Trans. Plasma Sci.* **2006**, *26*, 1685–1694. [CrossRef]
4. Winter, J.; Brandenburg, R.; Weltmann, K.D. Atmospheric pressure plasma jets: An overview of devices and new directions. *Plasma Sources Sci. Technol.* **2015**, *24*, 064001. [CrossRef]
5. Bruggeman, P.J.; Iza, F.; Brandenburg, R. Foundations of atmospheric pressure non-equilibrium plasmas. *Plasma Sources Sci. Technol.* **2017**, *26*, 123002. [CrossRef]
6. Tendero, C.; Tixier, C.; Tristant, P.; Desmaison, J.; Leprince, P. Atmospheric pressure plasmas: A review. *Spectrochim. Acta Part B At. Spectrosc.* **2006**, *21*, 2–30. [CrossRef]

7. Park, G.Y.; Park, S.J.; Choi, M.Y.; Koo, I.G.; Byun, J.H.; Hong, J.W.; Sim, J.Y.; Collins, G.J.; Lee, J.K. Atmospheric-pressure plasma sources for biomedical applications. *Plasma Sources Sci. Technol.* **2012**, *21*, 043001. [CrossRef]

8. Hoffmann, C.; Berganza, C.; Zhang, J. Cold atmospheric plasma: Methods of production and application in dentistry and oncology. *Med. Gas Res.* **2013**, *3*, 21. [CrossRef]

9. Fridman, G.; Friedman, G.; Gutsol, A.; Shekhter, A.; Vasilets, V.; Fridman, A. Applied plasma medicine. *Plasma Process. Polym.* **2008**, *5*, 503–533. [CrossRef]

10. Laroussi, M. Plasma medicine: A brief introduction. *Plasma* **2018**, *1*, 47–60. [CrossRef]

11. Morfill, G.; Kong, M.; Zimmermann, J. Focus on plasma medicine. *New J. Phys.* **2009**, *11*, 115011. [CrossRef]

12. Kong, M.G.; Kroesen, G.; Morfill, G.; Nosenko, T.; Shimizu, T.; Van Dijk, J.; Zimmermann, J.L. Plasma medicine: An introductory review. *New J. Phys.* **2009**, *11*, 115012. [CrossRef]

13. Tanaka, H.; Hori, M. Medical applications of non-thermal atmospheric pressure plasma. *J. Clin. Biochem. Nutr.* **2017**, *60*, 29–32. [CrossRef] [PubMed]

14. Choi, J.W.; Kang, S.U.; Kim, Y.E.; Park, J.K.; Yang, S.S.; Kim, Y.S.; Lee, Y.S.; Lee, Y.; Kim, C.-H. Novel therapeutic effects of non-thermal atmospheric pressure plasma for muscle regeneration and differentiation. *Sci. Rep.* **2016**, *6*, 28829. [CrossRef] [PubMed]

15. Chernets, N.; Zhang, J.; Steinbeck, M.J.; Kurpad, D.S.; Koyama, E.; Friedman, G.; Freeman, T.A. Nonthermal atmospheric pressure plasma enhances mouse limb bud survival, growth, and elongation. *Tissue Eng. Part A* **2015**, *21*, 300–309. [CrossRef] [PubMed]

16. Wang, M.; Cheng, X.; Zhu, W.; Holmes, B.; Keidar, M.; Zhang, L.G. Design of biomimetic and bioactive cold plasma-modified nanostructured scaffolds for enhanced osteogenic differentiation of bone marrow-derived mesenchymal stem cells. *Tissue Eng. Part A* **2013**, *20*, 1060–1071. [CrossRef]

17. Haertel, B.; Hahnel, M.; Blackert, S.; Wende K.; von Woedtke, T.; Lindequist U. Surface molecules on HaCaT keratinocytes after interaction with non-thermal atmospheric pressure plasma. *Cell Biol. Int.* **2012**, *36*, 1217–1222. [CrossRef]

18. Panngom, K.; Baik, K.Y.; Nam, M.K.; Han, J.H.; Rhim, H.; Choi, E.H. Preferential killing of human lung cancer cell lines with mitochondrial dysfuction by nonthermal dielectric barrier discharge plasma. *Cell Death Dis.* **2013**, *4*, e642. [CrossRef]

19. Schmidt, A.; Dietrich, S.; Steuer, A.; Weltmann, K.D.; von Woedtke, T.; Masur, K.; Wende, K. Non-thermal plasma activates human keratinocytes by stimulation of antioxidant and phase II pathways. *J. Biol. Chem.* **2015**, *290*, 6731–6750. [CrossRef]

20. Wende, K.; Straßenburg, S.; Haertel, B.; Harms, M.; Holtz, S.; Barton, A.; Masur, K.; von Woedtke, T.; Lindequist, U. Atmospheric pressure plasma jet treatment evokes transient oxidative stress in HaCaT keratinocytes and influences cell physiology. *Cell Biol. Int.* **2014**, *38*, 412–425. [CrossRef]

21. Georgescu, N.; Lupu, A.R. Tumoral and normal cells treatment with high-voltage pulsed cold atmospheric plasma jets. *IEEE Trans. Plasma Sci.* **2010**, *38*, 1949–1955. [CrossRef]

22. Finkel, T.; Holbrook, N.J. Oxidants, oxidative stress and the biology of ageing. *Nature* **2000**, *10*, 142–149. [CrossRef] [PubMed]

23. Lambeth, J.D. NOX enzymes and the biology of reactive oxygen. *Nat. Rev. Immunol.* **2004**, *4*, 181–189. [CrossRef] [PubMed]

24. Valko, M.; Leibfritz, D.; Moncol, J.; Cronin, M.T.; Mazur, M.; Telser, J. Free radicals and antioxidants in normal physiological functions and human disease. *Int. J. Biochem. Cell Biol.* **2007**, *39*, 44–84. [CrossRef] [PubMed]

25. Kim, S.J.; Chung, T.H. Cold atmospheric plasma-jet generated RONS and their selective effects on normal and carcinoma cells. *Sci. Rep.* **2008**, *6*, 20332. [CrossRef] [PubMed]

26. Fridman, G.; Shereshevsky, A.; Jostet, M.M. Floating electrode dielectric barrier discharge plasma in air promoting apoptotic behavior in melanoma skin cancer cell lines. *Plasma Chem. Plasma Process.* **2007**, *27*, 163–176. [CrossRef]

27. Barekzi, N.; Laroussi, M. Dose-dependent killing of leukemia cells by low-temperature plasma. *J. Phys. D Appl. Phys.* **2012**, *45*, 422002. [CrossRef]

28. Kalghatgi, S.U.; Fridman, G.; Cooper, M.; Nagaraj, G.; Peddinghaus, M.; Balasubramanian, M.; Vasilets, V.N.; Gutsol, A.F.; Fridman, A.; Friedman, G. Mechanism of blood coagulation by nonthermal atmospheric pressure dielectric barrier discharge plasma. *IEEE Trans. Plasma Sci.* **2007**, *35*, 1559–1566. [CrossRef]

29. Ma, R.N.; Feng, H.Q.; Liang, Y.D.; Zhang, Q.; Tian, Y.; Su, B.; Zhang, J.; Fang, J. An atmospheric-pressure cold plasma leads to apoptosis in *Saccharomyces cerevisiae* by accumulating intracellular reactive oxygen species and calcium. *J. Phys. D Appl. Phys.* **2013**, *46*, 285401–285408. [CrossRef]

30. Keidar, M.; Walk, R.; Shashurin, A.; Srinivasan, P.; Sandler, A.; Dasgupta, S.; Ravi, R.; Guerrero-Preston, R.; Trink, B. Cold plasma selectivity and the possibility of a paradigm shift in cancer therapy. *Br. J. Cancer* **2011**, *105*, 1295. [CrossRef]

31. Lee, H.J.; Shon, C.H.; Kim, Y.S.; Kim, S.; Kim, G.C.; Kong, M.G. Degradation of adhesion molecules of G361 melanoma cells by a non-thermal atmospheric pressure microplasma. *New J. Phys.* **2009**, *11*, 115026. [CrossRef]

32. Kim, G.C.; Kim, G.J.; Park, S.R.; Jeon, S.M.; Seo, H.J.; Iza, F.; Lee, J.K. Air plasma coupled with antibody-conjugated nanoparticles: A new weapon against cancer. *J. Phys. D* **2008**, *10*, 032005. [CrossRef]

33. Rivie, A.; Martus, K.; Menon, J. Atmospheric pressure plasma accelerates tail regeneration in tadpoles *Xenopus laevis*. *J. Eur. Phys. J. Spec. Top.* **2017**, *226*, 2859–2871. [CrossRef]

34. Haertel, B.; Von Woedtke, T.; Weltmann, K.D.; Lindequist, U. Non-thermal atmospheric-pressure plasma possible application in wound healing. *Biomol. Ther.* **2014**, *22*, 477–490. [CrossRef] [PubMed]

35. Xu, G.M.; Shi, X.M.; Cai, J.F.; Chen, S.L.; Li, P.; Yao, C.W.; Chang, Z.S.; Zhang, G.J. Dual effects of atmospheric pressure plasma jet on skin wound healing of mice. *Wound Repair Regen.* **2015**, *23*, 878–884. [CrossRef] [PubMed]

36. Fathollah, S.; Mirpour, S.; Mansouri, P.; Dehpour, A.R.; Ghoranneviss, M.; Rahimi, N.; Safaie Naraghi, Z.; Chalangari, R.; Chalangari, K.M. Investigation on the effects of the atmospheric pressure plasma on wound healing in diabetic rats. *Sci. Rep.* **2016**, *6*, 19144. [CrossRef] [PubMed]

37. McCallion, R.L.; Ferguson, M.W.J. Fetal wound healing and the development of antiscarring therapies for adult wound healing. In *The Molecular and Cellular Biology of Wound Repair*; Clark, R., Ed.; Springer: Boston, MA, USA, 1988; pp. 561–600, ISBN 978-1-4899-0185-9.

38. Dudas, M.; Wysocki, A.; Gelpi, B.; Tuan, T.L. Memory encoded throughout our bodies: Molecular and cellular basis of tissue regeneration. *Pediatr. Res.* **2008**, *63*, 502–512. [CrossRef]

39. Takeo, M.; Lee, W.; Ito, M. Wound healing and skin regeneration. *Cold Spring Harb. Perspect. Med.* **2015**, *5*, a023267. [CrossRef]

40. Bielefeld, K.A.; Amini-Nik, S.; Alman, B.A. Cutaneous wound healing: Recruiting developmental pathways for regeneration. *Cell. Mol. Life Sci.* **2013**, *10*, 2059–2081. [CrossRef]

41. Beck, C.W.; Christen, B.; Slack, J.M. Beyond early development: *Xenopus* as an emerging model for the study of regenerative mechanisms. *Dev. Dyn.* **2009**, *238*, 1226–1248. [CrossRef]

42. Demarquoy, J.; Le Borgne, F. Crosstalk between mitochondria and peroxisomes. *World J. Biol. Chem.* **2015**, *6*, 301–309. [CrossRef] [PubMed]

43. Schrader, M.; Yoon, Y.T. Mitochondria and peroxisomes: Are the "Big Brother" and the "Little Sister" closer than assumed? *BioEssays* **2007**, *29*, 1105–1114. [CrossRef] [PubMed]

44. Matés, J.M.; Segura, J.A.; Alonso, F.J.; Márquez, J. Oxidative stress in apoptosis and cancer: An update. *Arch. Toxicol.* **2012**, *11*, 1649–1665. [CrossRef] [PubMed]

45. Bonekamp, N.A.; Völkl, A.; Fahimi, H.D.; Schrader, M. Reactive oxygen species and peroxisomes: Struggling for balance. *BioFactors* **2009**, *35*, 346–355. [CrossRef] [PubMed]

46. Andreyev, A.Y.; Kushnareva, Y.E.; Starkov, A.A. Mitochondrial metabolism of reactive oxygen species. *Biochemistry* **2005**, *70*, 200–214. [CrossRef] [PubMed]

47. Turrens, J.F. Mitochondrial formation of reactive oxygen species. *J. Physiol.* **2003**, *552*, 335–344. [CrossRef] [PubMed]

48. Salas-Vidal, E.; Lomeli, H.; Castro-Obregon, S.; Cuervo, R.; Escalante-Alcalde, D.; Covarrubias, L. Reactive oxygen species participate in the control of mouse embryonic cell death. *Exp. Cell Res.* **1998**, *238*, 136–147. [CrossRef] [PubMed]

49. Duchen, M.R. Roles of mitochondria in health and disease. *Diabetes* **2004**, *53*, S96–S102. [CrossRef] [PubMed]

50. Baumgartner, H.K.; Gerasimenko, J.V.; Thorne, C.; Ferdek, P.; Pozzan, T.; Tepikin, A.V.; Petersen, O.H.; Sutton, R.; Watson, A.J.; Gerasimenko, O.V. Calcium elevation in mitochondria is the main Ca^{2+} requirement for mitochondrialpermeability transition pore (mPTP) opening. *J. Biol. Chem.* **2009**, *284*, 20796–20803. [CrossRef]

51. Bonora, M.; Wieckowski, M.R.; Chinopoulos, C.; Kepp, O.; Kroemer, G.; Galluzzi, L.; Pinton, P. Molecular mechanisms of cell death: Central implication of ATP synthase in mitochondrial permeability transition. *Oncogene* **2015**, *34*, 1475–1486. [CrossRef]

52. Morciano, G.; Giorgi, C.; Bonora, M.; Punzetti, S.; Pavasini, R.; Wieckowski, M.R.; Campo, G.; Pinton, P. Molecular identity of the mitochondrial permeability transition pore and its role in ischemia-reperfusion injury. *J. Mol. Cell. Cardiol.* **2015**, *78*, 142–153. [CrossRef] [PubMed]

53. Chen, G.; Wang, F.; Trachootham, D.; Huang, P. Preferential killing of cancer cells with mitochondrial dysfunction by natural compounds. *Mitochondrion* **2011**, *10*, 614–625. [CrossRef] [PubMed]

54. Zhivotovsky, B.; Orrenius, S. Calcium and cell death mechanisms. A perspective from the cell death community. *Cell Calcium* **2011**, *50*, 211–221. [CrossRef] [PubMed]

55. Pinton, P.; Giorgi, C.; Siviero, R.; Zecchini, E.; Rizzuto, R. Calcium and apoptosis: ER-mitochondria Ca^{2+} transfer in the control of apoptosis. *Oncogene* **2008**, *27*, 6407–6418. [CrossRef]

56. Yan, Y.; Wei, C.; Zhang, W.; Cheng, H.; Liu, J. Cross-talk between calcium and reactive oxygen species signaling. *Acta Pharmacol. Sin.* **2006**, *27*, 821–826. [CrossRef]

57. Burraco, P.; Valdés, A.E.; Johansson, F.; Gomez-Mestre, I. Physiological mechanisms of adaptive developmental plasticity in Rana temporaria island populations. *BMC Evol. Biol.* **2017**, *17*, 164. [CrossRef]

58. Gomez-Mestre, I.; Kulkarni, S.; Buchholz, D.R. Mechanisms and consequences of developmental acceleration in tadpoles responding to pond drying. *PLoS ONE* **2013**, *8*, e84266. [CrossRef]

59. Mangel, M.; Munch, S.B. A life-history perspective on short-and long-term consequences of compensatory growth. *Am. Nat.* **2005**, *166*, E155–E176. [CrossRef]

60. Slack, J.M.; Lin, G.; Chen, Y. The *Xenopus* tadpole: A new model for regeneration research. *Cell Mol. Life Sci.* **2008**, *65*, 54–63. [CrossRef]

61. Love, N.R.; Chen, Y.; Ishibashi, S.; Kritsiligkou, P.; Lea, R.; Koh, Y.; Gallop, J.L.; Dorey, K.; Amaya, E. Amputation-induced reactive oxygen species are required for successful *Xenopus* tadpole tail regeneration. *Nat. Cell Biol.* **2013**, *15*, 222–228. [CrossRef]

62. Beck, C.W.; Christen, B.; Barker, D.; Slack, J.M. Temporal requirement for bone morphogenetic proteins in regeneration of the tail and limb of *Xenopus* tadpoles. *Mech. Dev.* **2006**, *123*, 674–688. [CrossRef] [PubMed]

63. Nieuwkoop, P.; Faber, J. (Eds.) *Normal Table of Xenopus laevis (Daudin): A Systematic and Chronological Survey of the Development from the Fertilized Egg till the End of Metamorphosis*; North-Holland Publishing Company: Amsterdam, The Netherlands, 1967; ISBN 978-0815318965.

64. Denver, R.J.; Mirhadi, N.; Phillips, M. Adaptive plasticity in amphibian metamorphosis: Response of *Scaphiopus hammondii* tadpoles to habitat desiccation. *Ecology* **1998**, *79*, 1859–1872. [CrossRef]

65. De Block, M.; Stoks, R. Compensatory growth and oxidative stress in a damselfly. *Proc. R. Soc.* **2008**, *275*, 781–785. [CrossRef] [PubMed]

66. Denver, R.J. Proximate mechanisms of phenotypic plasticity in amphibian metamorphosis. *Integr. Comp. Biol.* **1997**, *37*, 172–184. [CrossRef]

67. Werner, E.E.; Anholt, B.R. Ecological consequences of the trade-off between growth and mortality rates mediated by foraging activity. *Am. Nat.* **1993**, *142*, 242–272. [CrossRef] [PubMed]

68. Wilbur, H.M.; Collins, J.P. Ecological aspects of amphibian metamorphosis. *Science* **1973**, *182*, 1305–1314. [CrossRef] [PubMed]

69. Echeverri, K.; Clarke, J.D.; Tanaka, E.M. In vivo imaging indicates muscle fiber dedifferentiation is a major contributor to the regenerating tail blastema. *Dev. Biol.* **2001**, *236*, 151–164. [CrossRef] [PubMed]

70. Schnapp, E.; Kragl, M.; Rubin, L.; Tanaka, E.M. Hedgehog signaling controls dorsoventral patterning, blastema cell proliferation and cartilage induction during axolotl tail regeneration. *Development* **2005**, *132*, 3243–3253. [CrossRef] [PubMed]

71. Whited, J.L.; Tabin, C.J. Regeneration review reprise. *J. Biol.* **2010**, *9*, 15. [CrossRef]

72. Gargioli, C.; Slack, J.M. Cell lineage tracing during *Xenopus* tail regeneration. *Development* **2004**, *131*, 2669–2679. [CrossRef]

73. Maki, N.; Suetsugu-Maki, R.; Tarui, H.; Agata, K.; Del Rio-Tsonis, K.; Tsonis, P.A. Expression of stem cell pluripotency factors during regeneration in newts. *Dev. Dyn.* **2009**, *238*, 1613–1616. [CrossRef] [PubMed]

74. Love, N.R.; Chen, Y.; Bonev, B.; Gilchrist, M.J.; Fairclough, L.; Lea, R.; Mohun, T.J.; Paredes, R.; Zeef, L.A.; Amaya, E. Genome-wide analysis of gene expression during *Xenopus tropicalis* tadpole tail regeneration. *BMC Dev. Biol.* **2011**, *10*, 70. [CrossRef] [PubMed]

75. Niethammer, P.; Grabher, C.; Look, A.T.; Mitchison, T.J. A tissue-scale gradient of hydrogen peroxide mediates rapid wound detection in zebrafish. *Nature* **2009**, *459*, 996–999. [CrossRef] [PubMed]

76. Gauron, C.; Rampon, C.; Bouzaffour, M.; Ipendey, E.; Teillon, J.; Volovitch, M.; Vriz, S. Sustained production of ROS triggers compensatory proliferation and is required for regeneration to proceed. *Sci. Rep.* **2013**, *3*, 2084. [CrossRef] [PubMed]

77. Hennings, H.; Michael, D.; Cheng, C.; Steinert, P.; Holbrook, K.; Yuspa, S.H. Calcium regulation of growth and differentiation of mouse epidermal cells in culture. *Cell* **1980**, *19*, 245–254. [CrossRef]

78. Mattson, M.P. Apoptosis in neurodegenerative disorders. *Nat. Rev. Mol. Cell. Biol.* **2000**, *1*, 120–129. [CrossRef] [PubMed]

79. Chan, D.C. Fusion and fission: Interlinked processes critical for mitochondrial health. *Ann. Rev. Genet.* **2012**, *46*, 265–287. [CrossRef] [PubMed]

80. Twig, G.; Shirihai, O.S. The interplay between mitochondrial dynamics and mitophagy. *Antioxid. Redox Signal* **2011**, *14*, 1939–1951. [CrossRef] [PubMed]

81. Szabadkai, G.; Simoni, A.M.; Chami, M.; Wieckowski, M.R.; Youle, R.J.; Rizzuto, R. Drp-1-dependent division of the mitochondrial network blocks intraorganellar Ca^{2+} waves and protects against Ca^{2+}-mediated apoptosis. *Mol. Cell* **2004**, *16*, 59–68. [CrossRef]

82. Youle, R.J.; Van Der Bliek, A.M. Mitochondrial fission, fusion, and stress. *Science* **2012**, *337*, 1062–1065. [CrossRef] [PubMed]

83. Rouzier, C.; Bannwarth, S.; Chaussenot, A.; Chevrollier, A.; Verschueren, A.; Bonello-Palot, N.; Fragaki, K.; Cano, A.; Pouget, J.; Pellissier, J.F.; et al. The *MFN2* gene is responsible for mitochondrial DNA instability and optic atrophy "plus" phenotype. *Brain* **2012**, *135*, 23–34. [CrossRef] [PubMed]

84. McCarron, J.G.; Wilson, C.; Sandison, M.E.; Olson, M.L.; Girkin, J.M.; Saunter, C.; Chalmers, S. From structure to function: Mitochondrial morphology, motion and shaping in vascular smooth muscle. *J. Vasc. Res.* **2013**, *50*, 357–371. [CrossRef] [PubMed]

85. Solenski, N.J.; diPierro, C.G.; Trimmer, P.A.; Kwan, A.L.; Helms, G.A. Ultrastructural changes of neuronal mitochondria after transient and permanent cerebral ischemia. *Stroke* **2002**, *33*, 816–824. [CrossRef] [PubMed]

86. Srinivasan, B. Mitochondrial permeability transition pore: An enigmatic gatekeeper. *New Horiz. Sci. Technol.* **1015**, *1*, 47–51.

87. Fridman, J.S.; Lowe, S.W. Control of apoptosis by p53. *Oncogene* **2003**, *22*, 9030–9040. [CrossRef]

88. van der Zand, A.; Gent, J.; Braakman, I.; Tabak, H.F. Biochemically distinct vesicles from the endoplasmic reticulum fuse to form peroxisomes. *Cell* **2012**, *149*, 397–409. [CrossRef] [PubMed]

89. Corpas, F.J.; Barroso, J.B.; del Río, L.A. Peroxisomes as a source of reactive oxygen species and nitric oxide signal molecules in plant cells. *Trends Plant Sci.* **2001**, *6*, 145–150. [CrossRef]

90. Furuse, M.; Hata, M.; Furuse, K.; Yoshida, Y.; Haratake, A.; Sugitani, Y.; Noda, T.; Kubo, A.; Tsukita, S. Claudin-based tight junctions are crucial for the mammalian epidermal barrier: A lesson from claudin-1-deficient mice. *J. Cell Biol.* **2002**, *156*, 1099–1111. [CrossRef]

91. Tattersall, G.J.; Wright, P.A. The effects of ambient pH on nitrogen excretion in early life stages of the American toad (*Bufo americanus*). *Comp. Biochem. Physiol. A Physiol.* **1996**, *113*, 369–374. [CrossRef]

92. Willens, S.; Stoskopf, M.K.; Baynes, R.E.; Lewbart, G.A.; Taylor, S.K.; Kennedy-Stoskopf, S. Percutaneous malathion absorption by anuran skin in flow-through diffusion cells. *Environ. Toxicol. Pharmacol.* **2006**, *22*, 255–262. [CrossRef]

93. Sumigray, K.D.; Lechler, T. Cell adhesion in epidermal development and barrier formation. In *Current Topics in Developmental Biology*; Yap, A., Ed.; Academic Press: Cambridge, MA, USA, 2015; pp. 383–414. [CrossRef]

 *applied sciences*

Article

Dispersion of OH Radicals in Applications Related to Fear-Free Dentistry Using Cold Plasma

Mehrdad Shahmohammadi Beni [1], Wei Han [2,3] and K.N. Yu [1,*

[1]   Department of Physics, City University of Hong Kong, Tat Chee Avenue, Kowloon Tong, Hong Kong, China; mshahmoha2-c@my.cityu.edu.hk
[2]   Center of Medical Physics and Technology, Hefei Institutes of Physical Sciences, Chinese Academy of Sciences, Hefei 230031, China; hanw@hfcas.ac.cn
[3]   Collaborative Innovation Center of Radiation Medicine of Jiangsu Higher Education Institutions and School for Radiological and Interdisciplinary Sciences (RAD-X), Soochow University, Suzhou 215168, China
*   Correspondence: peter.yu@cityu.edu.hk; Tel.: +852-34427182; Fax: +852-34420538

Received: 24 April 2019; Accepted: 20 May 2019; Published: 24 May 2019

Abstract: Cold atmospheric plasmas (CAPs) are being used in applications related to dentistry. Potential benefits include tooth whitening/bleaching, the sterilization of dental cavities, and root canal disinfection. Generated reactive species, such as hydroxyl (OH) radicals, play a critical role in the effectiveness of CAPs in dentistry. In the present work, the mandibular jaw and teeth were modeled. The propagation of CAP plume in ambient air was dynamically tracked using the level set method. The transport and dispersion OH radicals away from the nozzle and towards the teeth under treatment were also tracked. The distributions of concentration of OH radicals over the teeth were obtained for nozzle to tooth distances of 2 and 4 mm. The discharge of the OH radicals out of the nozzle was found to be asymmetrical. Interestingly, depending on the type of tooth treated, the dispersion of OH radicals out of the nozzle could be altered. The present model and obtained results could be useful for advancements towards a fear-free dentistry using CAPs.

Keywords: cold atmospheric plasmas; plasma medicine; dentistry; tooth whitening; fear-free dentistry

1. Introduction

In 1879, the fourth state of matter was unveiled by Sir William Crookes and subsequently in 1929 named "plasma" by Irving Langmuir [1,2]. Moreover, considering the phase transition (e.g., gas to plasma), there was a suggestion that plasmas not be called the fourth state of matter [3]. Plasma research has evolved rapidly and expanded in various areas including environmental, military, agricultural, and biomedical research [4,5]. Cold atmospheric plasmas (CAPs) enhance the generation of reactive species in their flow while the corresponding carrier gases remain at room temperature; this enhances the interaction of the reactive species with the target upon exposure of the latter to CAPs [6]. CAPs were also found useful in surface treatment, material processing, and plasma medicine [7–23]. In the field of medicine/biomedicine, CAPs are found to have significant potential in wound healing [24,25], cancer therapy [26–28], the inactivation of microorganisms [29], and tooth bleaching [21]. In particular, in the dentistry industry, CAPs were found effective against infected dental root canals and could remove microorganisms associated with infection of the root canal [30]. Various studies reported the effective killing of various types of viruses and bacteria by CAPs [31–35]. Considering the favorable characteristics of the CAPs, Stoffels et al. [33,36] were among the pioneers who explored the potential therapeutic use of CAPs in dentistry.

In dentistry, CAP applications were related to tooth whitening/bleaching, sterilization of dental cavities, and root canal disinfection. In a recent review [37], the applications of CAPs in dentistry were reviewed. Interestingly, it was pointed out that CAP treatment of teeth was a new and painless method for the preparation of cavities for restoration, with a rather enhanced longevity, root canal disinfection, and bacterial inactivation. In particular, tooth whitening using CAPs were experimentally studied. For example, Sun et al. [38] deployed a direct current air CAP to enhance tooth whitening. In a separate study, Lee et al. [21] found that a helium CAP jet enhanced tooth bleaching. The combined use of hydrogen peroxide (H_2O_2) with CAPs was found to be three times more effective to enhance tooth bleaching, when compared with the case without CAP. In another study, 40 extracted human molar teeth were treated with carbamide peroxide and additionally exposed to CAP, a plasma arc lamp, and a diode laser, which revealed that exposure to CAPs achieved the most effective tooth bleaching [12]. In addition, Claiborne et al. [39] applied CAP alongside H_2O_2 gel to demonstrate enhanced tooth whitening in 10- and 20-min exposure groups. Interestingly, CAPs were shown to be able to treat irregular surfaces such as dental cavities without the need of mechanical drilling [1]. As CAPs operate at room temperature, they would not exert detrimental effects on the tissues [33]. As regards root canal disinfection, Lu et al. [40] successfully inactivated *Enterococcus faecalis* bacterium in the root canal using a CAP jet device. Shali et al. [41] also examined the effectiveness of CAPs in the disinfection of the tooth root canal. From the progress in the use of CAPs in dentistry, it was hoped that CAPs could lead to fear-free and painless treatments in the near future.

Enhancement in plasma treatment of teeth was attributed to reactive species generated in CAPs, such as hydroxyl (OH) radicals. In the previous work of Pan et al. [42], teeth were exposed to saline solution with an air blow, 5% H_2O_2 gel and saline solution, and a CAP microjet. The authors found that the CAP microjet provided the best tooth whitening, which was due to the OH radicals as revealed using electron spin resonance spectroscopy [42]. In particular, minor enamel surface morphological changes were observed and a rougher surface was formed upon a CAP exposure for 20 min [42]. However, it was anticipated that longer CAP exposures could lead to more substantial morphological changes in the enamel, which might become unacceptable. On the other extreme, underexposures of the teeth to CAPs could jeopardize the efficacy of the treatment.

CAP devices with different designs could generate different distributions of reactive species over treated teeth [43], which could thus affect the treatment effectiveness. To the best of our knowledge, however, there were currently no standardized designs of CAP devices or theoretical models of CAP discharges and distribution of reactive species over teeth for dentistry. Such theoretical models, particularly in a time-dependent (i.e., dynamic) manner, would be critical in the success in applying CAPs to dentistry. Our group previously examined the interaction of CAPs with water medium [44], blood [45], and skin [46] using the finite element method (FEM). More recently, we studied the transport mechanism of OH radicals in CAP discharges and their dispersion and distribution over a skin layer [47], which considered both the diffusion and convection mechanisms in a two-phase flow system (i.e., CAP carrier gas and ambient air). Based on the success of previous models, in the present work, we analyzed the dispersion and distribution of OH radicals generated and distributed by CAPs over treated teeth. The present model could be useful for further development of CAP applications in dentistry, with the view to progress towards fear-free dentistry using CAPs.

2. Materials and Methods

The human mandibular jaw (i.e., the lower jaw) was modeled. Average dimensions of the main parts of the mandibular teeth were taken from previous measurements [48] and are shown in Table 1.

Table 1. Average dimensions of crown and root length of mandibular teeth.

Mandibular Teeth	Crown Length (mm)	Root Length (mm)
Central incisor	8.80	12.6
Lateral incisor	9.40	13.5
Canine	11.0	15.9
First premolar	8.80	14.4
Second premolar	8.20	14.7
First molar	7.70	14.0
Second molar	7.70	13.9
Third molar	7.50	11.8

The modeled mandibular jaw with teeth is shown in Figure 1, in which CAP discharge from a nozzle is also schematically shown. For modeling purposes, the mandibular jaw was placed in a rectangular chamber with the dimensions $90 \times 90 \times 45$ mm^3. The chamber was filled with air and all surfaces were set to be outlets (i.e., exposed to open air). The nozzle was assumed to be filled with helium (He) gas (i.e., the CAP carrier gas). The air in the chamber was assumed to be initially at rest and the plasma carrier gas was injected into the air domain at a specific flow rate Q. The plasma carrier gas (He) was discharged out of the nozzle with a radius of 1 mm and with a flowrate of 1 L/min ($Q = 1$ L/min) [47]. Following our previous work [47], the propagation of the carrier gas (He) in the air chamber was described using the level set method, since the flow resembled a two-phase flow system. The level set function was

$$\frac{\partial \phi}{\partial t} + \boldsymbol{u} \cdot \nabla\phi = \gamma\nabla\cdot\left(\varepsilon\nabla\phi - \phi(1-\phi)\frac{\nabla\phi}{|\nabla\phi|}\right) \tag{1}$$

where ϕ was the level set variable, $\boldsymbol{u}$ was the velocity field, t was time and γ was introduced for an enhanced numerical stability [49]. The level set function (i.e., Equation (1)) was fully coupled with Equation (2) which described the diffusion and convection of OH radicals:

$$\frac{\partial c_{OH}}{\partial t} + \nabla\cdot(-D_{OH}\nabla c_{OH}) + \boldsymbol{u}\cdot\nabla c_{OH} = S_{OH} + R_{OH} \tag{2}$$

where c_{OH} was the concentration of OH radicals, D was the diffusion coefficient, $\boldsymbol{u}$ was the velocity field, S_{OH} and R_{OH} were the source and reaction terms of the OH radicals [47]. The fluxes of OH radicals were defined in the nozzle to ensure the flow of these species from the nozzle along the axis perpendicular to the target (tooth under CAP treatment). The largest inflow concentration of OH radicals (i.e., at the OH source) was set to be 1 mol/m^3, which simplified the task in obtaining the OH concentrations in a normalized fashion over the target. In addition, variations in concentrations of OH radicals as a result of their reactions were modeled through a decay rate formulation. Interested readers are referred to our previous work [47] for more modeling details and for benchmarking with previous CAP discharge experiments. While OH radicals contributed to the bleaching effect, they might not be the main effectors. Short-lived reactive species in plasma-air-water system were analyzed [50] and different reactive species formed in CAPs were found to be responsible for tooth bleaching. In conventional dental bleaching, H_2O_2 was used (and still is used) as a bleaching agent. It is formed as a result of the recombination of OH radicals and via other reactions. Therefore, the transportation and dispersion of H_2O_2 species were also considered in our computational model in addition to the OH radicals. The diffusion coefficient data for H_2O_2 was taken from the previous work of Tang et al. [51].

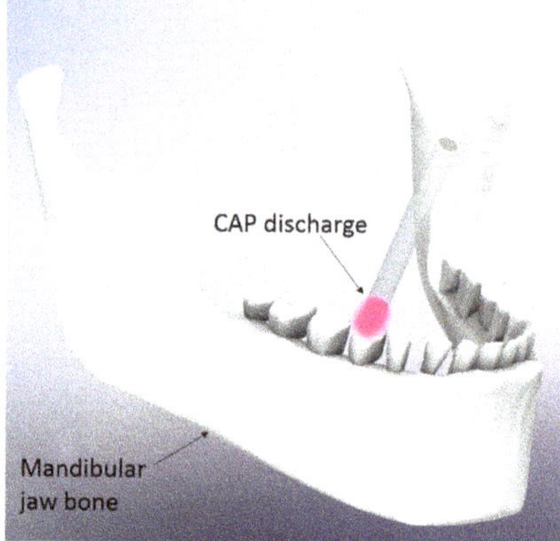

Figure 1. Modeled mandibular jaw with teeth under cold atmospheric plasma (CAP) discharge.

The CAP discharge tube was set to be perpendicular to the tooth under treatment, and the distance between the tube and the tooth was set to be 2 or 4 mm. The eight teeth on one side of the mandibular jaw were, starting from the center, the central and lateral incisor, canine, first and second premolar and first, second and third molar teeth, and eight teeth on the other side, which formed a mirror image of the previous eight teeth. In the present model, the geometry of the incisor and canine teeth were not significantly different from each another, hence, we studied the CAP exposure of six different teeth, namely, canine, first and second premolar and first, and second and third molar teeth. These teeth were numbered from 1 to 6 as shown in Figure 2.

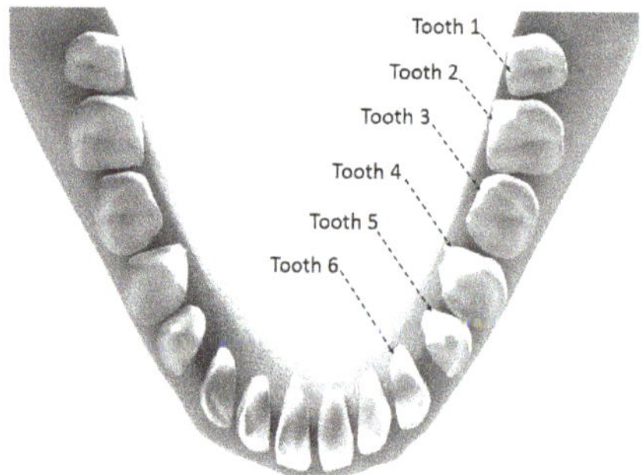

Figure 2. Schematic diagram showing teeth 1 to 6 examined in the present study.

The mandibular jaw bone was not the target of CAP exposures, and was therefore not considered during the computations, which helped improve the computational efficiency (including the total computation time and the required computational resources). The present numerical model was solved in parallel on a computer cluster with Intel Xeon E5-2630 v3 2.40 GHz processors (Intel Corporation, Santa Clara, CA, USA). The system was solved for 10 ms with time steps of 0.01 ms. It would be possible to solve the system using longer time steps; however, it might prevent the model from converging to the final solution effectively. The use of the finest time step was highly recommended, subject to the availability of computational resources.

3. Results and Discussion

During CAP discharge towards a treated tooth, the OH radicals were transported through diffusion and convection from the outlet of the nozzle towards the surface of the tooth. The OH radicals were distributed over the tooth, which led to a concentration gradient. The spatial variations of the concentration of the OH radicals over teeth 1 to 6 (see Figure 2) for the nozzle to tooth distance of 2 mm were obtained and shown in Figure 3. These snapshots were captured at $t = 10$ ms (i.e., the final timestep).

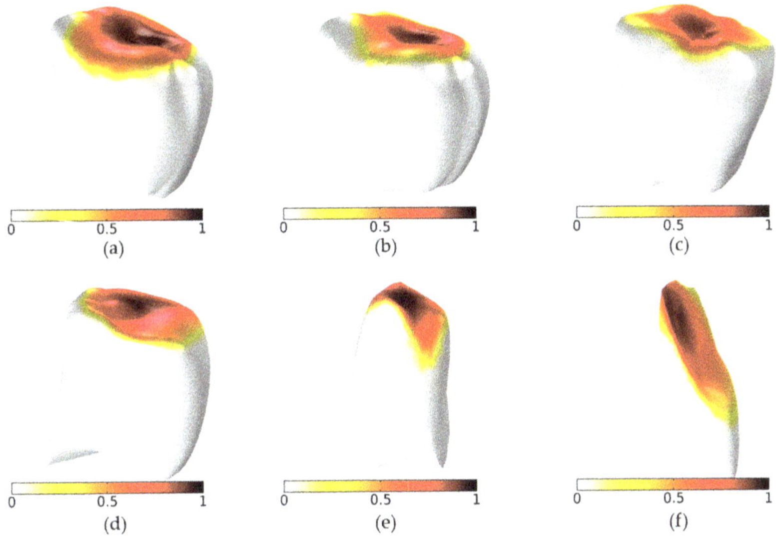

Figure 3. Spatial distribution of OH radicals over tooth (**a**) 1, (**b**) 2, (**c**) 3, (**d**) 4, (**e**) 5 and (**f**) 6 for nozzle to tooth distance of 2 mm. (Note: the scale bar shows the normalized concentration of the OH radicals).

Variations of OH concentrations over the six different exposed teeth for nozzle to tooth distance of 4 mm are shown in Figure 4. These snapshots were captured at $t = 10$ ms (i.e., the final timestep). To allow easier comparisons, orientations of the teeth were kept the same for both of the nozzle to tooth distances of 2 and 4 mm. Figures 3 and 4 showed that the OH concentration decreased, i.e., fewer OH radicals interacted with the tooth, as the nozzle to tooth distance increased. For longer nozzle to tooth distances, the area on a treated tooth exposed to OH radicals became smaller. Figure 5 also shows that fewer OH radicals could reach the treated tooth for longer nozzle to tooth distances if the speed and time duration were kept constant. Similarly, the variation of H_2O_2 species over the six different exposed teeth for nozzle to tooth distances of 2 and 4 mm are shown in Figures 5 and 6, respectively. The concentrations of H_2O_2 over the teeth decreased with the increasing nozzle to tooth distance, which was explained by the smaller amount of H_2O_2 covering the surfaces of exposed teeth for larger distances. These snapshots were also captured at $t = 10$ ms (i.e., the final timestep).

The normalized concentration of OH radicals over the tooth started to increase with time upon contact between the CAP plume carrying OH radicals and the tooth. The temporal variations of the normalized concentration of OH radicals averaged over the tooth varied for different teeth, due to their distinct geometries. For the nozzle to tooth distances of 2 and 4 mm, the concentrations of OH radicals started to increase for $t >$ ~1 ms (see Figure 7a) and for $t >$ ~2 ms (see Figure 7b), respectively. The OH radicals needed these onset time durations to travel from the nozzle to the tooth. The temporal variations of H_2O_2 species were similar to those of OH radicals. The H_2O_2 species were distributed over the surface of the exposed tooth and the normalized concentrations increased with time. The OH radicals had a comparatively larger diffusion coefficient compared to the H_2O_2 species [51,52], hence, the temporal variations of normalized concentrations of OH radicals were in general larger than those of the H_2O_2 species. The two main mechanisms for the dispersion of OH species (such as OH and H_2O_2) on different types of teeth were diffusion and convection, both of which were included in

Equation (2). More specifically, the reactive species diffused away from the plasma discharge through diffusion, and they were also carried to different types of teeth by the plasma carrier gas through convection. Figure 7a,b shows the temporal variation of normalized concentrations of OH and H_2O_2 species, respectively, which were averaged over a treated tooth surface. These results quantitatively described the dispersion of reactive species over the surface of the tooth under treatment, which was in turn related to the surface morphology of the tooth. For example, the results for tooth 6 (canine tooth) were significantly different from other molar and premolar teeth, which was attributed to the much sharper tip and smaller size of tooth 6.

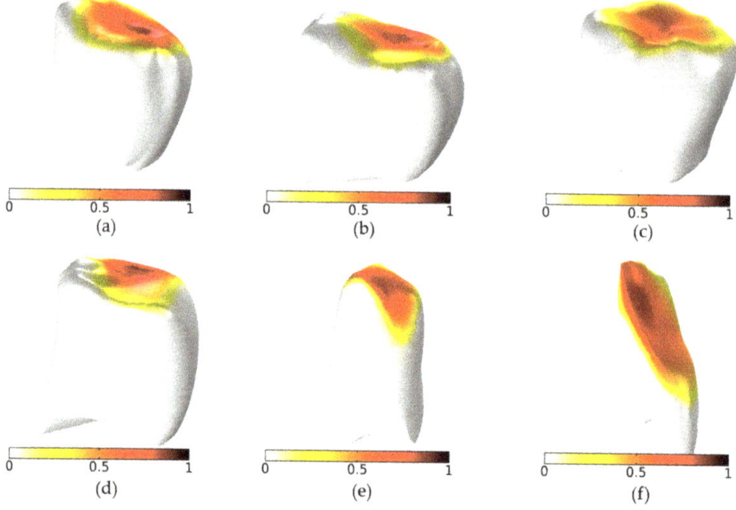

Figure 4. Spatial distribution of OH radicals over tooth (**a**) 1, (**b**) 2, (**c**) 3, (**d**) 4, (**e**) 5 and (**f**) 6 for nozzle to tooth distance of 4 mm. (Note: the scale bar shows the normalized concentration of the OH radicals).

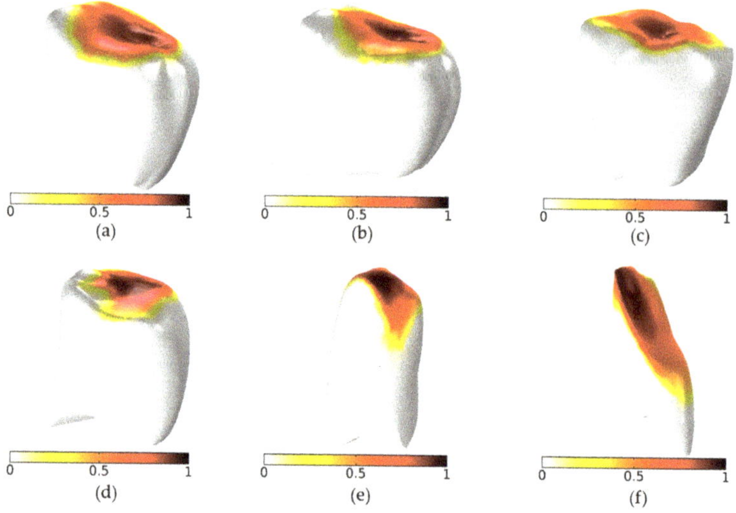

Figure 5. Spatial distribution of H_2O_2 species over tooth (**a**) 1, (**b**) 2, (**c**) 3, (**d**) 4, (**e**) 5 and (**f**) 6 for nozzle to tooth distance of 2 mm. (Note: the scale bar shows the normalized concentration of the H_2O_2 species).

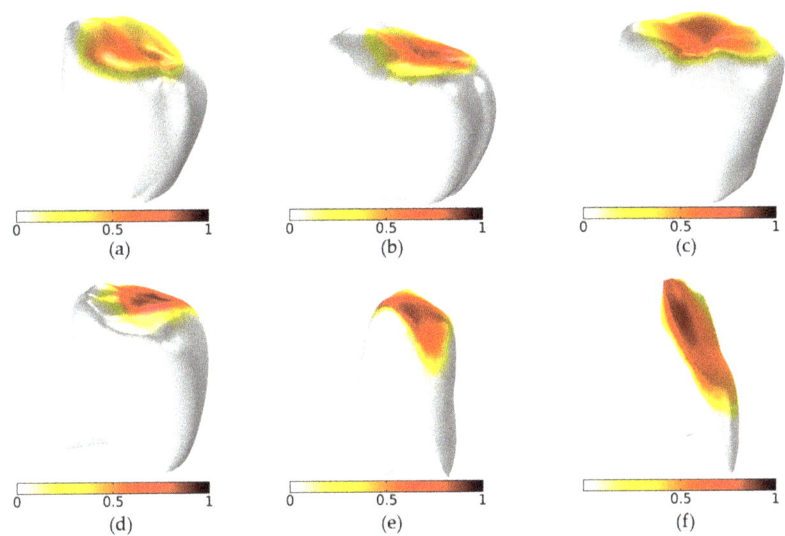

Figure 6. Spatial distribution of H_2O_2 species over tooth (**a**) 1, (**b**) 2, (**c**) 3, (**d**) 4, (**e**) 5 and (**f**) 6 for nozzle to tooth distance of 4 mm. (Note: the scale bar shows the normalized concentration of the H_2O_2 species).

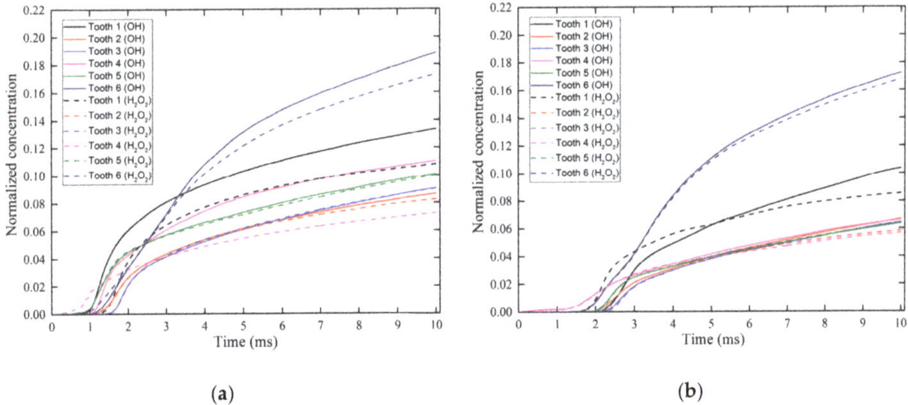

Figure 7. Temporal variations of normalized concentration of OH radicals and H_2O_2 species averaged over a treated tooth for nozzle to tooth distance of (**a**) 2 and (**b**) 4 mm.

To ascertain the steady-state in the studied system, the speed at 1 mm below the center of the nozzle was separately determined for all six teeth. The temporal variations of speeds at 1 mm below the center of the nozzle are shown in Figure 8a,b for 2 and 4 mm nozzle to tooth distances, respectively. The speed started to increase with time up to 1 ms, and became constant for the longer exposure time, which demonstrated the steady-state in the studied system. Moreover, the determined speeds varied for different treated teeth. Following the contact between the carrier gas and the tooth, the flow would get distorted and a velocity field would be generated (i.e., eddies would be formed) around the tooth. The distribution of this velocity field in three-dimensional (3D) space depended on the geometry of the treated tooth (or in other words how the flow was distorted), and led to variations in the speed (the magnitude of velocity in 3D space) at a point from the tooth.

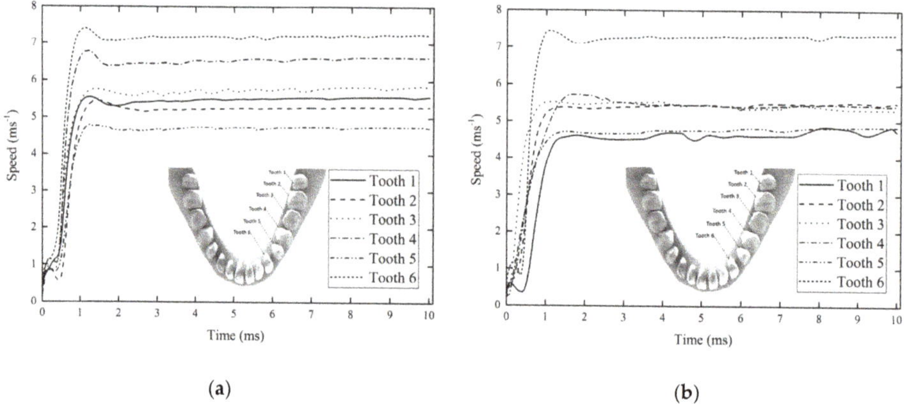

(a)　　　　　　　　　　　　　　　　　　　　**(b)**

Figure 8. Temporal variations of speed at 1 mm below the center of the nozzle for nozzle to tooth distance of (**a**) 2 and (**b**) 4 mm.

Figure 9a,b shows variations of normalized concentrations of OH radicals at 1 mm below the center of the nozzle for all six teeth at nozzle to tooth distances of 2 and 4 mm, respectively. The normalized concentrations started to increase rapidly up to about 1 in ~1 ms, beyond which the steady-state condition was achieved. Small variations for different teeth were noticed, which were attributed to the distortion of the flow and variations in the velocity field around different teeth under CAP treatment. Such variations were governed by the convection of the OH radicals, since the velocity field determined the dispersion and distribution of these species, which was mathematically shown in Equation (2), where the velocity field u was present. The results in Figure 9 imply that the movement of the OH radicals reached the steady-state as well.

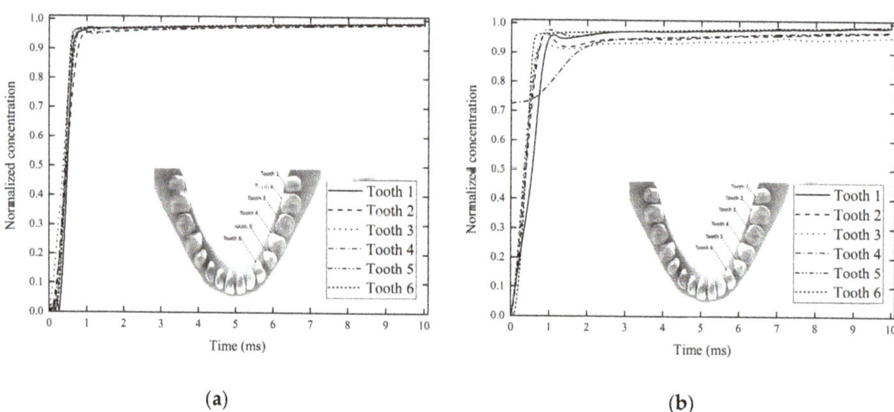

(a)　　　　　　　　　　　　　　　　　　　　**(b)**

Figure 9. Variations of normalized concentration of OH radicals at 1 mm below the center of the nozzle for nozzle to tooth distances of (**a**) 2 and (**b**) 4 mm.

Figure 10 shows the normalized concentrations of the OH radicals around the four sides at 1 mm below the nozzle, which were obtained for the nozzle to tooth distance of 2 mm. The results revealed anisotropy of dispersion and distribution of OH radicals underneath the nozzle and variations of the dispersion for different teeth. Interestingly, the normalized concentrations away from the center of the nozzle were lower than 1, which meant that most discharged OH radicals were concentrated at the center of the CAP plume. This feature was particularly useful considering that very narrow regions in teeth were often treated in dentistry. For $t < \sim1$ ms, the normalized concentrations significantly varied with time. In contrast, after the system reached the steady-state (i.e., $t > \sim 1$ ms), temporal variations in the normalized concentrations of OH radicals became negligible.

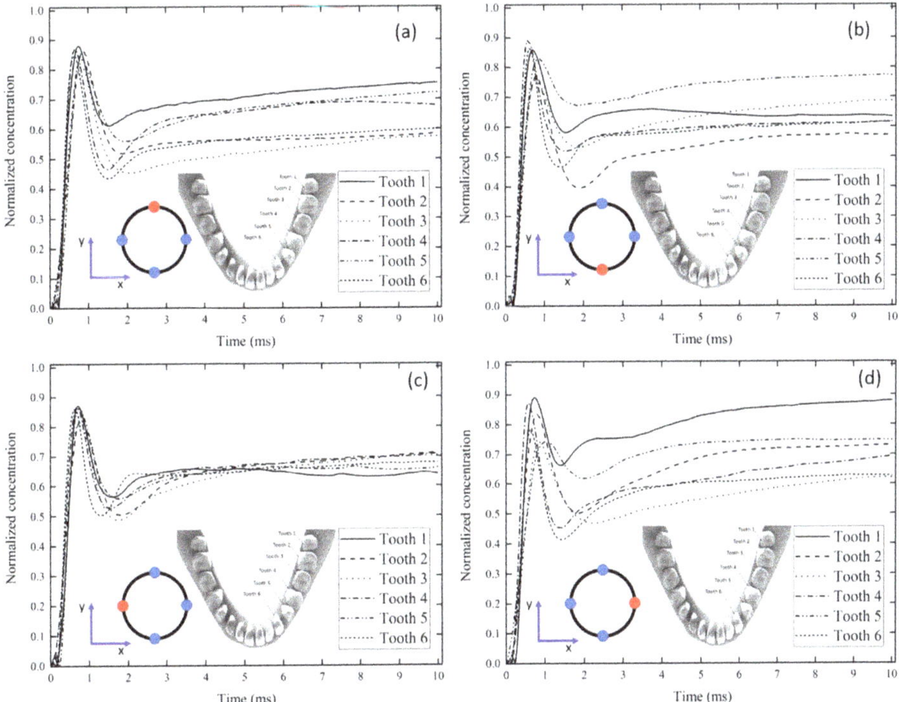

Figure 10. Variations of normalized concentrations of OH radicals at 1 mm below the nozzle at 1 mm (**a**) above, (**b**) below, (**c**) left, and (**d**) right of the center of the nozzle (from bird's eye view), for nozzle to tooth distance of 2 mm. (Inset: red dot indicates the location where the normalized concentration of OH was measured).

Similarly, Figure 11 shows the normalized concentrations of the OH radicals around the four sides at 1 mm below the nozzle, which were obtained for the nozzle to tooth distance of 4 mm. Again, temporal variations in the normalized concentrations of OH radicals were negligible for $t > \sim 1$ ms. Variations of normalized concentrations of OH radicals around the four sides of the nozzle were more considerable for the nozzle to tooth distance of 4 mm, which were attributed to the dispersion of OH radicals in the surrounding air before interacting with the tooth. For the longer nozzle to tooth distance of 4 mm, the OH radicals traveled longer distances to reach the tooth, which allowed some OH radicals to travel laterally away from the center of the flow through diffusion (i.e., $\nabla \cdot (-D_{OH} \nabla c_{OH})$ term in Equation (2)). Solving the present model in 3D space was in fact required due to anisotropic dispersion of the OH radicals.

In summary, Figures 10 and 11 highlighted the asymmetrical and sophisticated nature of dispersion and discharge of OH radicals out of the nozzle, which depended on the nozzle to tooth distance, as well as the type of tooth under treatment. Using the present model, the different geometries of nozzles and discharge tubes could be modeled with a view to achieving the best output distribution of reactive species according to the treatment requirements. Information regarding the distribution of reactive species could also be obtained to provide quantitative comparisons among different designs of CAP devices in dentistry.

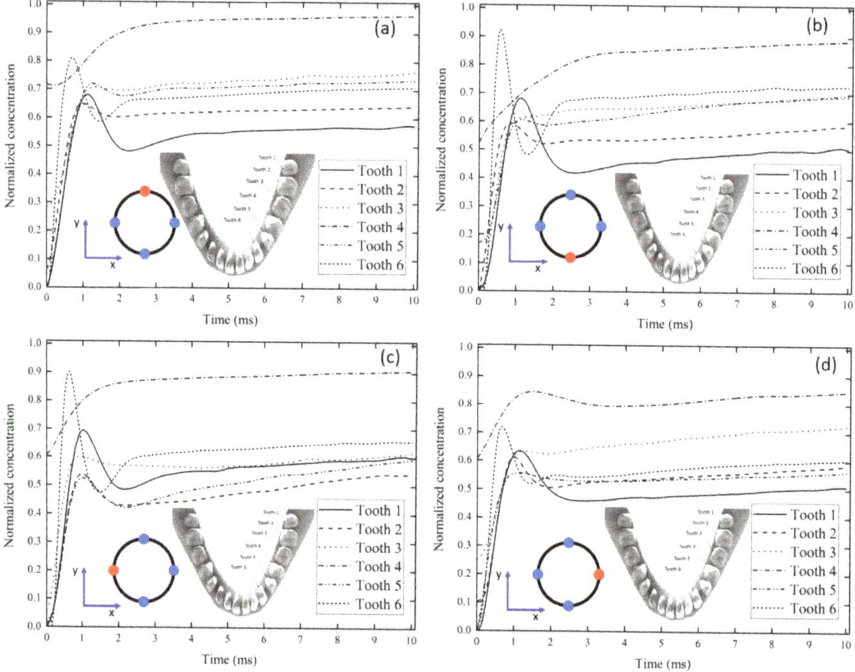

Figure 11. Variations of normalized concentrations of OH radicals at 1 mm below the nozzle at 1 mm (**a**) above, (**b**) below, (**c**) left, and (**d**) right of the center of the nozzle (from bird's eye view), for nozzle to tooth distance of 4 mm. (Inset: red dot indicates the location where the normalized concentration of OH was measured).

4. Conclusions

In the present work, a model was built for CAP discharges on teeth for dentistry applications for the first time. The model considered the CAP carrier gas and the ambient air through a two-phase flow model, and took into account the transport of OH radicals and H_2O_2 species involving diffusion and convection. The developed model determined the distributions of OH radicals and H_2O_2 species over six different chosen teeth. Different nozzle to tooth distances of 2 and 4 mm were examined and the obtained results were compared. The normalized concentration of OH radicals and H_2O_2 species over a treated tooth decreased with the increasing nozzle to tooth distance. The distribution of OH radicals out of the nozzle was found to be asymmetrical. The dispersion of OH radicals out of the nozzle varied according to the type of treated tooth. The dispersion of the reactive species became more significant for shorter distances between the nozzle and the tooth, which led to the coverage of a larger surface area of the tooth during plasma treatment. As such, the treatment could be more efficient. The developed model and the obtained results could be useful for working towards a fear-free dentistry using CAPs.

Author Contributions: Conceptualization, M.S.B.; methodology, M.S.B.; formal analysis, M.S.B., K.N.Y.; investigation, M.S.B.; resources, M.S.B., W.H., K.N.Y.; writing—original draft preparation, M.S.B.; writing—review and editing, W.H., K.N.Y.; visualization, M.S.B.; supervision, K.N.Y.; project administration, K.N.Y.; funding acquisition, W.H., K.N.Y.

Funding: This work was funded by National Natural Science Foundation of China (Nos. U1632145, 81573093 and 81227902), a project funded by the Priority Academic Program Development of Jiangsu Higher Education Institutions (PAPD) and Jiangsu Provincial Key Laboratory of Radiation Medicine and Protection, China Postdoctoral Science Foundation (No. 2016M592584).

Acknowledgments: We acknowledge the support of the Neutron computer cluster from the Department of Physics, City University of Hong Kong, for the computational work involved in this paper.

Conflicts of Interest: The authors declare no conflict of interest.

References

1. Kumar, S.C.; Sarada, P.; Reddy, S.C.; Reddy, S.M.; DSV, N. Plasma torch toothbrush a new insight in fear free dentisry. *J. Clin. Diagn. Res.* **2014**, *8*. [CrossRef]
2. Cha, S.; Park, Y.S. Plasma in dentistry. *Clin. Plasma Med.* **2014**, *2*, 4–10. [CrossRef] [PubMed]
3. Burm, K.T. Plasma: The fourth state of matter. *Plasma Chem. Plasma Process.* **2012**, *32*, 401–407. [CrossRef]
4. Heinlin, J.; Isbary, G.; Stolz, W.; Morfill, G.; Landthaler, M.; Shimizu, T.; Steffes, B.; Nosenko, T.; Zimmermann, J.L.; Karrer, S. Plasma applications in medicine with a special focus on dermatology. *J. Eur. Acad. Dermatol. Venereol.* **2011**, *25*, 1–11. [CrossRef]
5. Heinlin, J.; Morfill, G.; Landthaler, M.; Stolz, W.; Isbary, G.; Zimmermann, J.L.; Shimizu, T.; Karrer, S. Plasma medicine: Possible applications in dermatology. *J. Dtsch. Dermatol. Ges.* **2010**, *8*, 968–976. [CrossRef] [PubMed]
6. Pan, J.; Sun, K.; Liang, Y.; Sun, P.; Yang, X.; Wang, J.; Zhang, J.; Zhu, W.; Fang, J.; Becker, K.H. Cold plasma therapy of a tooth root canal infected with Enterococcus faecalis biofilms in vitro. *J. Endod.* **2013**, *39*, 105–110. [CrossRef]
7. Lu, X.; Naidis, G.V.; Laroussi, M.; Reuter, S.; Graves, D.B.; Ostrikov, K. Reactive species in non-equilibrium atmospheric-pressure plasmas: Generation, transport, and biological effects. *Phys. Rep.* **2016**, *630*. [CrossRef]
8. Lu, X.; Naidis, G.V.; Laroussi, M.; Ostrikov, K. Guided ionization waves: Theory and experiments. *Phys. Rep.* **2014**, *540*, 123–166. [CrossRef]
9. Graves, D.B. Low temperature plasma biomedicine: A tutorial review. *Phys. Plasmas* **2014**, *21*, 080901. [CrossRef]
10. Murakami, T.; Niemi, K.; Gans, T.; O'Connell, D.; Graham, W.G. Chemical kinetics and reactive species in atmospheric pressure helium–oxygen plasmas with humid-air impurities. *Plasma Sources Sci. Technol.* **2012**, *22*, 015003. [CrossRef]
11. Ji, L.; Bi, Z.; Niu, J.; Fan, H.; Liu, D. Atmospheric-pressure microplasmas with high current density confined inside helium-filled hollow-core fibers. *Appl. Phys. Lett.* **2013**, *102*, 184105. [CrossRef]
12. Nam, S.H.; Lee, H.W.; Cho, S.H.; Lee, J.K.; Jeon, Y.C.; Kim, G.C. High-efficiency tooth bleaching using non-thermal atmospheric pressure plasma with low concentration of hydrogen peroxide. *J. Appl. Oral Sci.* **2013**, *21*, 265–270. [CrossRef] [PubMed]
13. Foest, R.; Kindel, E.; Lange, H.; Ohl, A.; Stieber, M.; Weltmann, K.D. RF capillary jet-a tool for localized surface treatment. *Contrib. Plasma Phys.* **2007**, *47*, 119–128. [CrossRef]
14. Reuter, R.; Rügner, K.; Ellerweg, D.; de los Arcos, T.; von Keudell, A.; Benedikt, J. The role of oxygen and surface reactions in the deposition of silicon oxide like films from HMDSO at atmospheric pressure. *Plasma Process. Polym.* **2012**, *9*, 1116–1124. [CrossRef]
15. Murakami, T.; Niemi, K.; Gans, T.; O'Connell, D.; Graham, W.G. Afterglow chemistry of atmospheric-pressure helium–oxygen plasmas with humid air impurity. *Plasma Sources Sci. Technol.* **2014**, *23*, 025005. [CrossRef]
16. Naidis, G.V. Production of active species in cold helium–air plasma jets. *Plasma Sources Sci. Technol.* **2014**, *23*, 065014. [CrossRef]
17. Winter, J.; Brandenburg, R.; Weltmann, K.D. Atmospheric pressure plasma jets: An overview of devices and new directions. *Plasma Sources Sci. Technol.* **2015**, *24*, 064001. [CrossRef]
18. Flynn, P.B.; Busetti, A.; Wielogorska, E.; Chevallier, O.P.; Elliott, C.T.; Laverty, G.; Gorman, S.P.; Graham, W.G.; Gilmore, B.F. Non-thermal plasma exposure rapidly attenuates bacterial AHL-dependent quorum sensing and virulence. *Sci. Rep.* **2016**, *6*, 26320. [CrossRef] [PubMed]
19. Ikawa, S.; Kitano, K.; Hamaguchi, S. Effects of pH on bacterial inactivation in aqueous solutions due to low-temperature atmospheric pressure plasma application. *Plasma Process. Polym.* **2010**, *7*, 33–42. [CrossRef]
20. Laroussi, M. Low temperature plasma-based sterilization: Overview and state-of-the-art. *Plasma Process. Polym.* **2005**, *2*, 391–400. [CrossRef]
21. Lee, H.W.; Kim, G.J.; Kim, J.M.; Park, J.K.; Lee, J.K.; Kim, G.C. Tooth bleaching with nonthermal atmospheric pressure plasma. *J. Endod.* **2009**, *35*, 587–591. [CrossRef]

22. Laroussi, M. From killing bacteria to destroying cancer cells: 20 years of plasma medicine. *Plasma Process. Polym.* **2014**, *11*, 1138–1141. [CrossRef]

23. Kaushik, N.K.; Kim, Y.H.; Han, Y.G.; Choi, E.H. Effect of jet plasma on T98G human brain cancer cells. *Curr. Appl. Phys.* **2013**, *13*, 176–180. [CrossRef]

24. Isbary, G.; Morfill, G.; Schmidt, H.U.; Georgi, M.; Ramrath, K.; Heinlin, J.; Karrer, S.; Landthaler, M.; Shimizu, T.; Steffes, B.; et al. A first prospective randomized controlled trial to decrease bacterial load using cold atmospheric argon plasma on chronic wounds in patients. *Br. J. Dermatol.* **2010**, *163*, 78–82. [CrossRef]

25. Arndt, S.; Unger, P.; Wacker, E.; Shimizu, T.; Heinlin, J.; Li, Y.F.; Thomas, H.M.; Morfill, G.E.; Zimmermann, J.L.; Bosserhoff, A.K.; et al. Cold atmospheric plasma (CAP) changes gene expression of key molecules of the wound healing machinery and improves wound healing in vitro and in vivo. *PLoS ONE* **2013**, *8*, e79325. [CrossRef]

26. Cheng, X.; Murphy, W.; Recek, N.; Yan, D.; Cvelbar, U.; Vesel, A.; Mozetič, M.; Canady, J.; Keidar, M.; Sherman, J.H. Synergistic effect of gold nanoparticles and cold plasma on glioblastoma cancer therapy. *J. Phys. D Appl. Phys.* **2014**, *47*, 335402. [CrossRef]

27. Cheng, X.; Sherman, J.; Murphy, W.; Ratovitski, E.; Canady, J.; Keidar, M. The Effect of Tuning Cold Plasma Composition on Glioblastoma Cell Viability. *PLoS ONE* **2014**, *9*, e98652. [CrossRef]

28. Yan, D.; Talbot, A.; Nourmohammadi, N.; Cheng, X.; Canady, J.; Sherman, J.; Keidar, M. Principles of using cold atmospheric plasma stimulated media for cancer treatment. *Sci. Rep.* **2015**, *5*, 18339. [CrossRef]

29. Noriega, E.; Shama, G.; Laca, A.; Díaz, M.; Kong, M.G. Cold atmospheric gas plasma disinfection of chicken meat and chicken skin contaminated with Listeria innocua. *Food Microbiol.* **2011**, *28*, 1293–1300. [CrossRef]

30. Jiang, C.; Chen, M.T.; Gorur, A.; Schaudinn, C.; Jaramillo, D.E.; Costerton, J.W.; Sedghizadeh, P.P.; Vernier, P.T.; Gundersen, M.A. Nanosecond pulsed plasma dental probe. *Plasma Process. Polym.* **2009**, *6*, 479–483. [CrossRef]

31. Kuo, S.P.; Bivolaru, D.; Williams, S.; Carter, C.D. A microwave-augmented plasma torch module. *Plasma Sources Sci. Technol.* **2006**, *15*, 266. [CrossRef]

32. Tang, Y.Z.; Lu, X.P.; Laroussi, M.; Dobbs, F.C. Sublethal and killing effects of atmospheric-pressure, nonthermal plasma on eukaryotic microalgae in aqueous media. *Plasma Process. Polym.* **2008**, *5*, 552–558. [CrossRef]

33. Sladek, R.E.; Stoffels, E.; Walraven, R.; Tielbeek, P.J.; Koolhoven, R.A. Plasma treatment of dental cavities: A feasibility study. *IEEE Trans. Plasma Sci.* **2004**, *32*, 1540–1543. [CrossRef]

34. Stoffels, E.; Kieft, I.E.; Sladek, R.E.; Van den Bedem, L.J.; Van der Laan, E.P.; Steinbuch, M. Plasma needle for in vivo medical treatment: Recent developments and perspectives. *Plasma Sources Sci. Technol.* **2006**, *15*, S169. [CrossRef]

35. Sladek, R.E.; Filoche, S.K.; Sissons, C.H.; Stoffels, E. Treatment of Streptococcus mutans biofilms with a nonthermal atmospheric plasma. *Lett. Appl. Microbiol.* **2007**, *45*, 318–323. [CrossRef]

36. Stoffels, E.; Flikweert, A.J.; Stoffels, W.W.; Kroesen, G.M. Plasma needle: A non-destructive atmospheric plasma source for fine surface treatment of (bio) materials. *Plasma Sources Sci. Technol.* **2002**, *11*, 383. [CrossRef]

37. Arora, V.; Nikhil, V.; Suri, N.K.; Arora, P. Cold atmospheric plasma (CAP) in dentistry. *Dentistry* **2014**, *4*, 1. [CrossRef]

38. Sun, P.; Pan, J.; Tian, Y.; Bai, N.; Wu, H.; Wang, L.; Yu, C.; Zhang, J.; Zhu, W.; Becker, K.H.; et al. Tooth whitening with hydrogen peroxide assisted by a direct-current cold atmospheric-pressure air plasma microjet. *IEEE Trans. Plasma Sci.* **2010**, *38*, 1892–1896.

39. Claiborne, D.; McCombs, G.; Lemaster, M.; Akman, M.A.; Laroussi, M. Low-temperature atmospheric pressure plasma enhanced tooth whitening: The next-generation technology. *Int. J. Dent. Hyg.* **2014**, *12*, 108–114. [CrossRef]

40. Lu, X.; Cao, Y.; Yang, P.; Xiong, Q.; Xiong, Z.; Xian, Y.; Pan, Y. An RC plasma device for sterilization of root canal of teeth. *IEEE Trans. Plasma Sci.* **2009**, *37*, 668–673.

41. Shali, P.; Asadi, P.; Ashari, M.A.; Shokri, B. Cold atmospheric pressure plasma jet for tooth root canal disinfection. In Proceedings of the IEEE International Conference on Plasma Sciences (ICOPS), Antalya, Turkey, 24–28 May 2015.

42. Pan, J.; Sun, P.; Tian, Y.; Zhou, H.; Wu, H.; Bai, N.; Liu, F.; Zhu, W.; Zhang, J.; Becker, K.H.; et al. A novel method of tooth whitening using cold plasma microjet driven by direct current in atmospheric-pressure air. *IEEE Trans. Plasma Sci.* **2010**, *38*, 3143–3151. [CrossRef]

43. Yue, Y.; Pei, X.; Lu, X. Comparison on the absolute concentrations of hydroxyl and atomic oxygen generated by five different nonequilibrium atmospheric-pressure plasma jets. *IEEE Trans. Radiat. Plasma Med. Sci.* **2017**, *1*, 541–549. [CrossRef]

44. Shahmohammadi Beni, M.; Yu, K.N. Computational fluid dynamics analysis of cold plasma carrier gas injected into a fluid using level set method. *Biointerphases* **2015**, *10*, 041003. [CrossRef]

45. Shahmohammadi Beni, M.; Yu, K.N. Computational Fluid Dynamics Analysis of Cold Plasma Plume Mixing with Blood Using Level Set Method Coupled with Heat Transfer. *Appl. Sci.* **2017**, *7*, 578. [CrossRef]

46. Shahmohammadi Beni, M.; Yu, K.N. Safeguarding against inactivation temperatures during plasma treatment of skin: Multiphysics model and phase field method. *Math. Comput. Appl.* **2017**, *22*, 24. [CrossRef]

47. Shahmohammadi Beni, M.; Han, W.; Yu, K.N. Modeling OH transport phenomena in cold plasma discharges using level set method. *Plasma Sci. Technol.* **2019**, *21*, 055403. [CrossRef]

48. Scheid, R.C. *Woelfel's Dental Anatomy*; Lippincott Williams & Wilkins: Philadelphia, PA, USA, 2012.

49. Shahmohammadi Beni, M.; Zhao, J.; Yu, K.N. Investigation of droplet behaviors for spray cooling using level set method. *Ann. Nucl. Energy* **2018**, *113*, 162–170. [CrossRef]

50. Gorbanev, Y.; Privat-Maldonado, A.; Bogaerts, A. Analysis of Short-Lived Reactive Species in Plasma–Air–Water Systems: The Dos and the Do Nots. *Anal. Chem.* **2018**, *90*, 13151–13158. [CrossRef] [PubMed]

51. Tang, M.J.; Cox, R.A.; Kalberer, M. Compilation and evaluation of gas phase diffusion coefficients of reactive trace gases in the atmosphere: Volume 1. Inorganic compounds. *Atmos. Chem. Phys.* **2014**, *14*, 9233–9247. [CrossRef]

52. Liu, Y.; Ivanov, A.V.; Molina, M.J. Temperature dependence of OH diffusion in air and He. *Geophys. Res. Lett.* **2009**, *36*, L03816. [CrossRef]

MDPI
St. Alban-Anlage 66
4052 Basel
Switzerland
Tel. +41 61 683 77 34
Fax +41 61 302 89 18
www.mdpi.com

Applied Sciences Editorial Office
E-mail: applsci@mdpi.com
www.mdpi.com/journal/applsci